To Paul
with good memories
of some strange weeks
1994,

Ingrid

T H E
BHOPAL
S A G A

Causes and Consequences of the
World's Largest Industrial Disaster

Ingrid Eckerman

Universities Press

Universities Press (India) Private Limited

Registered Office
3-5-819 Hyderguda, Hyderabad 500 029 (A.P.), India
Email: hyd2_upilco@sancharnet.in

Distributed by
Orient Longman Private Limited

Registered Office
3-6-752 Himayatnagar, Hyderabad 500 029 (A.P.), India

Other Offices
Bangalore / Bhopal / Bhubaneshwar / Chennai
Ernakulam / Guwahati / Hyderabad / Jaipur / Kolkata
Lucknow / Mumbai / New Delhi / Patna

© Universities Press (India) Private Limited 2005

First published 2005

ISBN 81 7371 515 7

Typeset by
OSDATA, Hyderabad 500 029

Printed in India at
Orion Printers Private Limited, Hyderabad 500 004

Published by
Universities Press (India) Private Limited
3-5-819 Hyderguda, Hyderabad 500 029

CONTENTS

PREFACE

So clearly do I still remember the scenes on television on December 3, 1984, as well as my own reactions: " Of course, it was a multinational that caused this catastrophe. Of course, they will deny causing it. Of course, it was a very poor and powerless population that was hit. Of course, they will not get the support to which they should have the right."

Nearly ten years later in 1994, I visited Bhopal for the first time, as a member of the International Medical Commission on Bhopal. I felt greatly honoured to be welcomed by survivors from the television scenes, with flowers, food and speeches.

The submissions I received during this first stay verified that my reactions in 1984 were correct.

Since 1994, I have been back every year and have followed the survivors' and activists' work for justice and support.

I found that hundreds or even thousands of articles have been published as well as several reports and books. Yet, all this material confused me, as it was sometimes contradictory. It was not clear to me what had really happened, what was the cause and what were the consequences. The role of Union Carbide as well as other actors was not clear. What were rumours and tales, and where was the truth? I had a need to clarify this for myself.

As a family physician, a general doctor, I could not limit myself. I wanted to find out everything about the catastrophe. Planning for a Master of Public Health, I started to read, systematise and structure all information I got my hands on. It was not possible to include all material and all discussions in the essay, so the essay became a book.

In 1999, the manuscript was sent to Union Carbide for scrutiny, but there was no response.

I dedicate this book to all those in Bhopal who are now my friends.

Eckerman, Ingrid
Stockholm, Sweden, January 2004
E-mail: ingrid.eckerman@slpo.sll.se/
eckerman@algonet.se

ABBREVIATIONS AND INDIAN TERMS

2.1 Abbreviations

BGIA	Bhopal Group for Information and Action
BGPMUS	Bhopal Gas Peedit Mahila Udyog Sangathan, a survivors' organisation
BMHRC	Bhopal Memorial Hospital and Research Centre
BMHT	Bhopal Memorial Hospital Trust
CBI	Central Bureau of Investigation
CSIR	Council for Scientific and Industrial Research
ICMR	Indian Council of Medical Research
IMCB	International Medical Commission on Bhopal
MIC	Methyl Isocyanate
MP	Madhya Pradesh, an Indian state
NGO	Non-governmental organisation, voluntary association
Rs	Rupees, the Indian currency. 40 rupees is about 1 US$ (2003)
UCC	Union Carbide Corporation
UCIL	Union Carbide India Limited

2.2 Indian terms

Crore	10 millions
Lakh	100,000
Dharna	Sit-in

Basti	Settlement
Kuccha	Roughly-built (house, toilet etc)
Pucca	Permanent, built of concrete (house, toilet etc)
Coolie	Bearer
Bhurka	Chaperone, the body-covering coat of Muslim women

2.3 Technical terms

Psi	Pounds per square inch (pressure)
Psig	Pounds per square inch gauge
Ppm	Parts per million (concentration)

3

SUMMARY

Introduction

The Bhopal gas leak, India 1984, is the largest chemical industrial accident ever. An estimated 520,000 persons were exposed to the gases, and up to 8,000 died during the first weeks. 100,000 persons or more have suffered permanent injuries. The catastrophe has become the symbol of negligence of human beings on the part of transnational corporations. It has thus served as an alarm bell. All the same, industrial disasters still happen, in India, as well as in the industrialised part of the world. Although they are far from the size of Bhopal, they are so numerous that chemical hazards could well be considered as a public health problem. The companies usually dispute their own role in the accidents, and deny the health effects of the accidents. The companies have also been reluctant to compensate the victims economically.

In injury analysis, the concept "the process of the accident", including pre-event, event and post event phases, is used. Many models for injury analysis have been developed. Usually, they are used for events like traffic accidents and child burns.

Methods

The book is based on a thorough review of already published material from India and outside India, and the author's experiences from visiting the city of Bhopal. The Logical Framework Approach (LFA), a tool for project planning and management, has been tested on this mega accident, in order to analyse the causes and its consequences.

Results

The Logical Framework Approach (LFA) provides one main message: That irrespective of the direct cause of the leakage, it is only two parties that are responsible for the magnitude of the disaster: Union Carbide Corporation and the Governments of India and Madhya Pradesh.

LFA appears to be a complete and useful model for analysing a complex situation like the Bhopal gas leak. The problem and objectives trees look like a chain of events from where there are branches and roots. Despite a thorough knowledge of the Bhopal gas leak, developing this problem tree gave the author some new views on the connection between causes and effects. However, the tree looks more like a "problem net". Also, when drawing the tree of objectives, the author got some new ideas on measures necessary to prevent an accident or mitigate the effects of it. The matrix makes it possible to clarify what processes/changes from other instances are needed if the project is to succeed.

Conclusions

Models developed for analysis of injuries can be used for analysing a complicated mega accident like the Bhopal gas leak, although different models might stress different aspects. Visualising the causes and consequences in tree models might provide a new understanding. When visualising the causes and consequences of this kind of accident, it is obvious that "chain" or "tree" are not the right words. "Net" is more appropriate.

Analysis, according to the LFA problem tree, demonstrates that to create the mega-gas leak, it was not enough that water entered the tank. The most important factors were the plant design and the cutting down of expenses because of economic pressure.

The same analysis shows that the most important factor for the outcome of the leakage is the negligence of the Union Carbide Corporation and the Governments of India and Madhya Pradesh.

To reduce the influence of chemical industries on public health, there is a great need for actions from many actors. The governments have a responsibility to protect their inhabitants from the negative effects of "development". As a result of globalisation, co-ordination between governments and national organisations is necessary.

4

INTRODUCTION TO THE BHOPAL SAGA

4.1 The chemical industry and public health

During the last century, the chemical industry, including the pharmaceutical industry, has grown and developed enormously. It is estimated that several hundred new chemical compounds are being synthesised every day. We know very little about the effects of these compounds on human beings, animals and eco-systems, in either the short- or the long-term. The trial and error method seems to be the most common method used for risk assessment.

A number of these compounds, as well as many intermediates and waste products, are toxic to nature and to human beings. The health of workers and residents in proximity to mines, transportation routes and plants will be affected, as well as coming generations.

When we talk about "pesticides and developing countries", we should not only consider the use of pesticides, but also their production. Production of pesticides is a part of the chemical industry, which is growing rapidly in developing countries. The history of the chemical industry is replete with chemical accidents and the exposure of workers and people around the plants. The chemical industry is an important factor from a public health point of view.

Pesticides and the chemical industry in developing countries are a public health problem also for those of us who live in high-income countries. Hazardous products and processes exported to the third world are already returning to the developed countries in a "circle of poison". The physical consequences of a chemical accident might spread to other countries through water, wind and food. The political and economic consequences might also spread all over the world.

Cassels [1] points out that the internationalisation of both business and environmental degradation is increasingly teaching the developed world that the provision of aid to poorer countries – in order to better manage the risks and effects of industrial development – may no longer be a matter of charity, but an imperative motivated by self-interest.

The gas leakage from Union Carbide's plant in Bhopal, India, in 1984 is the largest industrial hazard ever experienced in the world. Over 500,000 persons were exposed to the gases; between 3,000 and 10,000 people died within the first weeks; and between 100,000 and 200,000 may have permanent injuries.

Because of its magnitude, this catastrophe has not been forgotten. Hundreds and hundreds of articles have been published as well as several books. Different kinds of research have been done. The process that led to the leakage, the effects of the gases, and the actions of the company, the government and the medical and scientific establishments have been documented in hundreds or thousands of articles and several reports and books.

There is no evidence that the processes and actions would have been different with other chemical accidents. On the contrary, the knowledge about Bhopal can be used when we study other accidents or discuss what measures should be taken to prevent exposure to toxic chemical substances.

The Bhopal gas leakage has become a symbol of transnational corporate negligence towards human beings. It has thus served as a wake-up call. All the same, industrial disasters still happen in India, as well as in the industrialised parts of the world. Although they are far from the size of Bhopal, they are so numerous that chemical accidents could well be considered as a public health problem. The companies usually dispute their own role in the accidents and deny the effects of the accidents on health. The companies have also been reluctant to compensate the victims economically.

In 1985, 40 workers had to be hospitalized because of a chlorine leak from a textile mill in Kerala. In Gujarat in 1987,

5,000 persons were injured because of a gas leakage. In an explosion in a ship in Maharashtra in 1991, 100 persons were killed. During 1994, around 50 chemical and/or fire accidents were voluntarily reported from within Madhya Pradesh to the Disaster Management Institute in Bhopal (personal correspondence).

Industrial disasters also happen in the industrialised part of the world. In the USA, during 1985, more than 16,000 persons were evacuated, 30 were killed and 31 injured in nine chemical accidents. In Union Carbide's factories in USA and Europe, more than 700 have died, several hundred have been injured and 17,000 have been evacuated because of accidents. After installing a new safety system at the MIC-plant at Institute, West Virginia, 135 people were injured by toxic gases in 1985. Wells and vegetables have been contaminated and ruined by Union Carbide pesticides.

The pharmaceutical industries are no better. In 1976 in Seveso, Italy, 1,000 persons were evacuated and 100,000 cattle died. During 1986, massive amounts of toxic chemicals, including 66,000 pounds of pesticides, were accidentally released into the river Rhine by Sandoz, Hoechst, Bayer and Ciba-Geigy. In 1997, during the building of railway tunnels in Sweden and Norway, the use of the tightening compound Rhoca-Gil, manufactured by Rhône-Poulenc, led to the leakage of acrylic amid into the groundwater.

There are some similarities between the accidents:

- The catastrophes affect countries outside those where the transnational companies are seated. Production is often established in countries where regulations are less stringent.
- Trade unions and occupational healthcare seem to have been poorly developed, having little influence on the work environment.
- It seems as though the catastrophes could have been predicted and prevented.
- The companies have disputed their own role in the accidents and denied the health effects of the accidents.
- The companies have been reluctant to compensate the victims economically.

"Syndrome" from a medical point of view is a combination of several symptoms and findings. So far, we have used "the Bhopal syndrome" for the combinations of symptoms and affected organs of the gas affected. But maybe, one could talk also about the "Bhopal syndrome" as a matter of occurrence. We know that small Bhopals happen every day – the mechanisms and the affected organs seem to be the same as for the original Bhopal catastrophe.

4.2 Human rights

After the Bhopal tragedy, the Permanent Peoples' Tribunal in Bhopal in 1992 concluded that fundamental human rights had been grossly violated in terms of a series of articles in the various international declarations concerned with human rights [2].

National governments have signed those declarations and many try seriously to follow them. Thus, it is possible to sue with the International Tribunal for Human Rights in the Hague.

However, sometimes transnational companies have more power than national governments. In 1993, the 15 largest corporations in the world had gross incomes greater than the gross domestic products of over 120 countries [3]. So far, these corporations have not signed any declarations on human rights.

One might expect that human rights would include the right to life and health as well as the right to a healthy environment.

However, in the *International Bill of Human Rights*, nothing is said directly about human rights to life and health, or human rights to a healthy environment [4].

In the *Universal Declaration of Human Rights*, the following text was found, which to some extent, supports the idea of rights to health and a healthy environment:

> "Everyone, as a member of society, has the right to social security and is entitled to realisation, through national effort and international co-operation and in accordance with the organisation and resources of each State, of the economic, social and cultural rights indispensable for his dignity and the free development of his personality" (Article 22).

"Everyone has the right to a standard of living adequate
for the health and well-being of himself and of his family,
including food, clothing, housing and medical care and
necessary social services, and the right to security in the
event of unemployment, sickness, disability, widowhood,
old age or other lack of livelihood in circumstances
beyond his control" (Article 25:1).

The *International Covenant on Civil and Political Rights* in
Article 6:1 states that "every human being has the inherent
right to life. This right shall be protected by law. No one shall
be arbitrarily deprived of his life."

The *International Covenant on Economic, Social and Cultural
Rights*, however, is clearer on rights, health and environment.
Article 8 talks about the right to form and join trade unions. In
Article 12, the "right of everyone to the enjoyment of the highest
attainable standard of physical and mental health" is
recognised. It includes child health, improvement of environmental and industrial hygiene, prevention and treatment of occupational diseases, and assurance of medical service and
medical attention.

4.3 Bhopal: History, geography and demography

Bhopal is an old town with a Mogul past. In the 1950s, it had
only around 60,000 inhabitants. When it became the capital of
Madhya Pradesh, it started to grow rapidly. It is the centre of administration, education and political and economic power as well
as culture. It is also the centre of communication and a railway
junction.

Bhopal is situated in the centre of India, 500 metres above sea
level. It is surrounded by hills, forests and fields. There are two
large lakes, the Upper Lake and the Lower Lake, which are actually two dams, built one thousand years ago. The area of the
Bhopal municipality covers 285 sq. km.

North of the dams lies the Old Town, with narrow streets,
2–4 storey houses, markets, mosques and the railway station. It is
crowded with people. This is where the poorer sections of the

population live. They are labourers, loaders, handcart pullers, carpenters and domestic servants. Many keep cattle. South of the dams is the New Town, with some parks, broad avenues, modern building complexes here and there and widespread villa quarters. Both areas are partially mixed with and surrounded by squatter slum areas, but by and large, fairly rich people live here.

Bhopal is surrounded by hills. The plant of Union Carbide India Limited (UCIL) is situated in the north, adjacent to densely populated slum wards and the railway station. As early as 1975, it was pointed out that 12 colonies thickly surrounded the factory and this posed serious environmental hazards [5]. Further south is the Old Town, which slopes upwards toward the Lakes and the New Town. New slum areas have grown around the factory, even on ground that is most probably still polluted by different chemicals from the factory [6].

In 1984, Bhopal had around 800,000 inhabitants. In 1981, 38% of the population were in the age group 0–14 years [7]. The 36 wards that were classified as affected had around 520,000 inhabitants [8]. This means that around 200,000 children between 0–14 years were affected by the gases.

In 1985, of the total affected, 80% earned below Rs 145 per month (around US$ 5). 1.2% earned more than Rs 465 per month; 47% lived in a kuccha (non-permanent) house; 50% were Hindus and 49% Muslims [8]. Muslims are generally considered to be poorer than Hindus and it is less common for Muslim women to work outside the home.

The total population of Bhopal in 1994 was 1,062,800. The number of houseless households was 5,331. With a mean household size of five people, this means that around 29,000 people do not have a house [9].

The density of the population is 3,746 per sq.km. For the city, the birth rate is 29 per 1,000, the crude death rate, 9 per 1,000 and the infant mortality rate, 78 per 1,000 live births [10].

After the gas disaster, it was found that 55% of the population in the Old City was Hindu and 43% Muslim [11].

The per capita expenditure on family welfare services by the Government of Madhya Pradesh was Rs 7.08 during 1985–1986 [10].

The inflation rate has been high. In 1994, 1 rupee was worth one-third (or 30 paisa) of the 1984 rate. Bhopal is said to have become one of the most expensive towns in India since the interim relief began to be paid out.

4.4 References

1. Cassels, J., *The Uncertain Promise of Law: Lessons from Bhopal*. 1993, Toronto: University of Toronto Press Inc.
2. *Asia '92. Permanent Peoples' Tribunal. Findings and Judgements.* in *Third session on industrial and environmental hazards and human rights*. 1992, Oct 19–24. Bhopal-Bombay.
3. Morehouse, W., *The ethics of industrial disaster in a transnational world: The elusive quest for justice and accountability in Bhopal*.
4. *The International Bill of Human Rights*. 1993, United Nations: New York.
5. *Socio-economic Impact of Disbursement of Interim Relief to Gas-affected Famillies of Bhopal*. 1991, Academy of Administration, Government of Madhya Pradesh: Bhopal. p. 154.
6. *Union Carbide in Bhopal, India. The Lingering Legacy. Analyses of carbide related toxins at the former UCIL site*. 1990, National Toxics Campaign Fund: Boston.
7. *Census 1981*. 1982, Government of Madhya Pradesh, The Census Department: Bhopal.
8. *Annual Report 1990*. 1990, Indian Council of Medical Research, Bhopal Gas Disaster Research Centre, Gandhi Medical College: Bhopal.
9. *Census 1991*. 1991, Government of Madhya Pradesh, The Census Department: Bhopal.
10. *Report on family welfare/primary health care needs of slumdwellers of Bhopal town and formulation of proposal for strengthening of existing family welfare services and creation of new health facilities*. 1992, Department of Preventive and Social Medicine, Gandhi Medical College: Bhopal.
11. *Compensation Disbursement. Problems and Possiblities*. 1992, Bhopal Group for Information and Action: Bhopal.

OBJECTIVES AND METHODS

5.1 Objectives

The general objective of this report is to produce a reliable overview of the gas leakage in Bhopal in 1984.

The specific objectives are to analyse available material to find answers to the following issues:

- If and how the accident could have been prevented;
- What the probable components of the gases were;
- Whether these components could explain the deaths, the permanent injuries and the distribution of symptoms and injuries from grave to light;
- If and how the immediate treatment of the survivors could have been more efficient;
- If and how the long-term effects on health could have been mitigated;
- If and how the long-term socio-economic effects could have been mitigated;
- If the accident had any influence on the safety policies of the chemical industry in India.

To use the knowledge gained in a discussion on:

- How chemical disasters can be prevented;
- How outbreak epidemiology should be designed;
- The demands to make of authorities, medical establishment and the WHO in the prevention and management of industrial disasters.

5.2 Methods

The book is based on material already published in India and other countries, and on the author's experiences from repeated visits to the city of Bhopal, totalling eight months during 10 years.

The collected material can be classified as follows.

Scientific papers and reports
- By members of IMCB
- By other (medical) authors
- Research on MIC

Material from ICMR
- Manuals and the like
- Annual reports
- Other reports

Material from official Indian authorities
- Government of Madhya Pradesh
- Central Bureau of Investigation
- Disaster Management Institute
- Gandhi Medical College
- Council for Scientific and Industrial Research

Material from Union Carbide
- Manuals published before 1984
- Pamphlets, video and other material published after 1984

Material published by NGOs
- BGIA: pamphlets
- Sambhavna: newsletters, annual reports, papers
- Other NGOs

Books
- *Bhopal: Industrial Genocide?* (Arena Press) A collection of 30 articles published in different newspapers in 1984.

- *Bhopal Gas Tragedy* (Delhi Science Forum). A thorough examination of the causes and the management of the accident, written in 1984 and 1985.
- *Nothing to Lose but Our Lives* (Dembo et al). A series of articles by different authors, dealing with industrial hazards.
- *Corporate Killings. Bhopals Will Happen* (Jones) is a thorough examination of what was known in 1987.
- *The Bhopal Syndrome. Pesticides, Environment and Health* (Weir).
- *The Uncertain Promise of Law: Lessons from Bhopal* (Cassels) is a thorough examination of the relations between multinational corporations, governments and the people.
- *Bhopal: The Inside Story* (Chouhan). The main section was written by a former MIC plant operator in the UCIL factory in Bhopal. One appendix has testimonies from 16 plant personnel. Another describes the legal issues.
- *The Ophidian and the Orphans of Bhopal* (Pandey) is mainly about legal issues.
- *Bhopal Tragedy. Socio-Legal Implications* (Chauhan) deals with the economic compensation and the social rehabilitation.
- *It was Five Past Midnight in Bhopal* (Lapierre, Moro) is a novel based on interviews and studied material.
- *Silent Invaders* (Jacobs et al) is a series of articles on pesticides and women's health.
- *Asphyxiating Asia* (Mac Sheoin) is a thorough examination of what the development of chemical industry in Asia means to people and environment.

Articles from non-scientific papers

Reports

Background material
- Toxicology and pathology
- Pathology
- Injuries
- Post traumatic stress disorder
- Night work
- Socio-economic conditions and health

- Environmental risk management
- Human rights

Web sites
Submissions (in addition to the material above)
- from survivors to IMCB
- from activists and doctors to IMCB

Interviews
- Ms Rashida Bee, president of the Bhopal Gas Affected Women's Stationery Workers' Union
- Ms Champa Devi Shukla, secretary of the Bhopal Gas Affected Women's Stationery Workers' Union
- Mr Abdul Jabbar Khan, convenor of the Bhopal Gas Peedit Mahila Udyog Sangathan
- Mr T R Chouhan, a former operator at the UCIL plant in Bhopal
- Mr N D Jayaprakash, Bhopal Gas Peedit Sangharsh Sahayog Samiti and Delhi Science Forum
- Mr Satinath Sarangi, convenor of the BGIA and managing director of Sambhavna clinic
- The late Dr Dwivedi, MD, former director of ICMR in Bhopal
- Mr Deena Deenadayalan, the Other Media, Delhi
- Drs Desphande, Qaiser and Kaur, Sambhavna Clinic

Many statements and submissions are repeated in different articles and books. For practical reasons, some books and articles have been chosen as major references. When the statements in other sources are congruent with the major references, they will not be indicated in the text.

The full list of reference material and other studied material is found at the end of this report.

The conceptual models for accident and injury analysis by Haddon and Berger were originally used to analyse the causes of the disaster and its consequences. As a complement, the Logical Framework Approach was tested on this mega-accident. As this model seems more complete and useful for this complex situation, it is included in this book. However, the Haddon structure pre-event, event and post-event phases, is used for the structure of the book.

5.3 Comments

The quality of the collected material can be a topic of discussion. This is especially true with regard to the submissions from survivors and workers.

- Submissions by survivors may have been biased because of hopes of economic gain.
- Submissions by workers may have been biased due to the intention of avoiding responsibility.
- Even when the survivors' mother tongue is Urdu, the interviews are likely to have been held in Hindi.
- Submissions have been translated into English; some qualities may have been lost.
- Foreign journalists generally do not know Urdu or Hindi. Many interviews must have gone through an interpreter or were conducted in bad English.
- Many of the original statements or submissions have been repeated in different articles and papers in a way that makes it seem as if they were new.

Quite a few times, data about the material is missing – author, publisher and/or year of publishing. Referring to research has been done, without references being given. This creates an uncertainty about the relevance and adequacy of the material.

Some material is missing, because it has not been released.
- It is very probable that UC knows much more about the composition of the cloud compared to the information they have released.
- It is likely that there still exists information from governmental institutions which has not been released, for instance, about why the Government of India suddenly accepted the sum of compensation offered by UC.
- ICMR's annual reports were kept secret for many years.
- The surveys undertaken by the Tata Institute immediately after the disaster have still not been released.

The quality of the epidemiological and clinical research material will be discussed in Chapter 8.

Although the list of studied material is extensive, I am fully aware that I have missed information. More interviews should have been done with doctors and officials in Bhopal. I hope those who have something more to add will contact me.

6

THE PRE-EVENT PHASE: THE PROCESS THAT LED TO THE LEAKAGE

6.1 Background

6.1.1 The Green Revolution

In the 1960s, the market for pesticides in Europe had started to become saturated and because of new knowledge and protests from environmental activists, it became restricted. The multinational corporations then turned to the Third World, which offered cheap labour, low maintenance costs and showed a relative indifference to occupational health.

During the 1950s and 60s, crop failures and famines were common occurrences in Asia. The "Green Revolution" was considered to be the solution, not only by the chemical industry, but also by farmers, governments and NGOs [1]. In India, the Green Revolution displaced traditional growing methods with high yielding seed varieties that required large amounts of fertilisers and pesticides. The government encouraged the production of pesticides locally, but it was insufficient. In 1966, Indian leaders decided to turn to foreign manufacturers. Union Carbide Corporation (UCC) immediately imported American Sevin, and undertook to build a Sevin-producing factory within five years in India.

The effects of pesticides on health soon became obvious. Illiterate peasants, labourers and their families were exposed to massive doses of the toxins when handling the pesticides, without receiving instructions or safety precautions [1]. Accidents were regularly reported in the newspapers. In some areas, cancer rates rose alarmingly. Psychological disorders became common. Pesticides became the most used method for committing suicide. Peasants who doubled or tripled the recommended dosage in the hope of doubling or tripling their production, were ruined.

India had isolated itself from the global economy [2]. At the time of the Green Revolution, India had greater autonomy in relation to multinational capital than had Latin American countries. The Green Revolution was a co-production of state and capital. At the time of the Bhopal massacre, the balance of power between state and capital had shifted, with the collapse of the command economies and India's desire for economic modernisation and liberalisation.

The Indian Government was very keen on establishing chemical industries. Indian authorities stated in 1998 that one of the competitive advantages that India had was that companies were comparatively free to pollute there [3].

6.1.2 Union Carbide Corporation

The creation in 1917 and the development of Union Carbide Corporation (UCC) are described by Mac Sheoin [2]. UCC has violated the environment and the health of workers and residents innumerable times [1, 2, 4–6]. For example, during the years 1980–1984, 67 leakages of methyl isocyanate (MIC) occurred at the Institute's factory in West Virginia. The management of the factory took care not to bring these leaks to the attention of the people living in the valley, claiming that none of them had posed a real threat to health, nor exceeded the legally accepted standards for toxic emissions in the atmosphere [1].

As part of its internationalisation, Union Carbide became India's first petrochemical producer. For nearly a century, UCC's lamps and batteries brought light to remote farms and villages [1]. In India, UCC manufactured chemical products, plastic goods, photographic plates, films, industrial electrodes, polyester resin, laminated glass and machine tools.

The Union Carbide factory, Bhopal in 1984 (Sambhavna Trust)

The Union Carbide logo (Sambhavna Trust)

Union Carbide Corporation (UCC) is the parent company, and Union Carbide India Limited (UCIL) the Indian subsidiary. While UCC held 50.9% of the stock, various Indian investors, including public sector financial institutions, held the rest.

UCC was allowed majority ownership, despite government limitations on foreign investment, because of the technological sophistication of its operations. UCC chose all production processes, supplied all plant designs and designated operational procedures. It also conducted the safety audits.

Managerial control over UCIL was exercised by Union Carbide through its Eastern Division headquarters in Hong Kong [7]. There is evidence that even minor production and maintenance decisions were made there [6–8].

6.1.3 The UCIL plant in Bhopal

The process of creating the UCIL plant in Bhopal is described by Lapierre and Moro [1]. It was welcomed not only by the authorities, but also by the residents, who saw opportunities to get jobs there. To work for UCIL meant a high salary as well as high social status. UCIL developed social activities like sport and entertainment that involved the whole community.

The Union Carbide plant in Bhopal opened in 1969 to manufacture pesticides [6, 7]. At first, UCIL in Bhopal only formulated carbamate pesticides from concentrates imported from the US.

The pesticide plant at Bhopal was a facility set up in 1977 for the manufacture of Sevin (carbaryl) and its formulations. Initially, the primary raw materials were imported from the USA. The highly reactive MIC was transported by lorries, with police escort, from Mumbai (Bombay) to Bhopal. Only from 1980, did the Bhopal plant start manufacturing MIC, using the know-how and designs supplied by the parent company in the USA. The plant also produced carbon monoxide and phosgene, both of which are required for the production of MIC [9–11].

The Government of India granted UCIL a licence to manufacture 5,000 tonnes of Sevin a year. The UCC man in Mumbai realised that annual sales would not exceed 2,000 tonnes. He tried in vain to persuade the management committee in New York to plan a smaller plant. The president, directors and engineers were obsessed with the idea of creating the most beautiful pesticide plant in India and ignored the advice from the man on the ground [1].

The production of pesticides at the UCIL plant in Bhopal was not a great success. Repeated droughts combined with the effectiveness of the pesticides being unsatisfactory, led to a decline in pesticide sales. In 1982, sales equalled less than half the production capacity – in 1984, less than a fifth. In 1982, the alpha-naphthol plant was closed down – it had never been able to supply a product that was pure enough [1, 12]. The MIC unit was over-dimensioned from the beginning and had always run at a loss. Because of this, the work force was reduced. Just prior to the accident, UCC had been considering ways in which it might divest itself of the Bhopal operation or dispose of the plant. One suggestion was to dismantle the plant and ship it to Brazil or Indonesia, but objections to this were raised "because of the high corrosion at several points" [1, 12].

The Bhopal plant included a research centre, the biggest in Asia, with five insect-rearing laboratories and a two-hectare experimental farm for testing chemical agents [1, 13, 14].

From 1977 to 1982, the pesticide plant was managed by an American engineer, who had one essential principle: "Always keep only a strict minimum of dangerous materials on site." He was also keen on safety precautions. He worked hard at being a good leader. In 1982, he was replaced by an Indian-born engineer, educated in the USA. The directors of UCIL made him subject to a financial controller whose purpose was to reduce the factory's losses. These two men had a more hierarchical sense. As wages constituted the primary expense, they started to dismiss first *coolies*, and later, skilled workers and technicians [1, 8].

6.2 Designs Important for Safety

Between 1958 and 1973, Union Carbide used an alternative way of producing carbamate pesticides, without using MIC. For economic reasons, this was changed to a more hazardous method [10].

MIC is produced by a reaction of phosgene and monomethyl-amine (MMA) into methylcarbamyl chloride (MCC) and hydrogen chloride (HCl). MCC is then pyrolysed to yield MIC and HCl [15]. Chloroform is used as a solvent in the MIC process, and caustic lye for the neutralisation of any toxic material that needs to be disposed of.

Chouhan [10], Delhi Science Forum [9], CSIR [15], and the Disaster Management Institute [11] describe the design of the MIC plant in Bhopal:

Instead of using a "closed loop" process, where MIC was converted as soon as it was manufactured, UCC applied for large-scale storage of MIC, consisting of three horizontally mounted stainless steel tanks. Normally two of the tanks (no. 610 and 611) were used to store the product when it was of acceptable quality. The third tank (no. 619) was used for temporary storage of off-specification material until it was reprocessed. The two tanks could hold around 90 tonnes of MIC, sufficient for about 30 days' production of Sevin. Any off-specification material not reprocessed was sent to the vent gas scrubber for neutralisation with caustic solution.

The UC manual [16] states that one tank should always be kept empty, that the tanks should not be filled to more than 60 percent capacity, that the tanks should be kept under an atmosphere of nitrogen with a pressure 1,0 kg/cm^2, and that the temperature should be kept below $+5°$ C. Only stainless steel tanks should be used; the use of iron, carbon steel, aluminium, zinc, galvanised iron, tin, copper and their alloys was prohibited.

The relieve valve vent header (RVVH) provided a relief line for toxic gases to be routed to the vent gas scrubber (VGS), in the event that a pressure build-up in any one of the tanks caused a large volume of gas to escape [8]. If the pressure within the tank were to exceed 40 psig[*], it would cause a rupture disc fitted to the end of the RVVH line to give. The gas so released would force the relief valve to open, which allowed the gas to flow down the RVVH directly to the VGS.

A second line, called the process vent header (PVH) led from the tanks to the VGS. The nitrogen pressurisation system was connected to this line. To ensure that MIC did not come into

---

[*] Pounds per square inch above atmospheric pressure

contact with moisture in the air, the chemical was stored under pressure and protected by a blanket of dry nitrogen. If the pressure in a tank fell below the operating pressure of 2 psig, nitrogen was fed into the tank [8].

The VGS consisted of columns packed with ceramic balls. The bottom portion of the scrubber held about 21,000 gallons (95 m^3) of 10 percent caustic solution, which was pumped to the top of the scrubber. The waste gases were scrubbed by a counter-current flow. After scrubbing, the gas was flared or released through a 33.5 m high stack. While the flare tower primarily burnt vent gases from the carbon monoxide unit, it was also used to burn vent gases from the MIC storage tanks.

The written instructions which the workers were to follow in washing out the lines at the MIC unit omitted the procedure that a slip-bind be inserted, according to a report in 1985 by the Union Research Group of Mumbai (Bombay), "The role of management practices in the Bhopal Gas Leak Disaster" [8].

Until May 1984, the lines to the RVVH and the PVH had been unconnected and each performed a separate function [8]. The management wanted a standby line in the event of either the PVH or the RVVH having to be shut down for repair. A jumper line between the two lines provided an easy solution.

When the plant was first designed, the managing director of UCIL recommended that the preliminary design of the Bhopal MIC facility be altered to involve only token storage in small individual containers. However, UCIL was overruled by the parent corporation, which insisted on a design similar to the West Virginia plant [1, 6, 17].

But for reasons of economy, Bhopal's "beautiful plant" would not be provided with all the safety equipment and security systems the engineers in South Charleston had envisaged. The precise reasons for these economies remain obscure [1].

In 1975, the state government prepared a master plan for the city of Bhopal. A sparsely populated site outside Bhopal was designated as an industrial area for hazardous facilities. In spite of warnings, Union Carbide insisted on building the MIC production and storage unit at an existing Union Carbide facility upwind from the city, mainly because it was cheaper to draw on the infrastructure of the existing facility [1, 18].

The chief engineer was worried about the squatters just outside the wall around the factory area. He asked the municipal authorities to force people to move, and proposed drawing up an evacuation plan. [1].

The policy of UCC was to maintain "centralised integrated corporate strategic planning, directing and control" [17]. The Indian subsidiary had to send detailed monthly operating reports to UCC [8]. All accidents involving fatal or serious injuries "will be reviewed by the UCC executive officer". In 1982, over 30 percent of UCIL's raw materials, spare parts and components were imported [13]. Training for the Bhopal management was also provided in the USA. UCIL was one of the few firms in India in which the parent company was allowed to maintain a majority interest, because it possessed all the "know-how".

After the leakage, UCC's first line of defence was that the equipment installed in Bhopal was made in the USA to US specifications, with safety equipment and standards virtually identical in both Bhopal and the Institute, Virginia [13]. Because of the reactions in Virginia, UCC was forced to later admit that this was not the case.

Survivors and NGOs have accused UCC of using double standards for safety when planning factories in developing countries. In Table 1, we can compare the design used for the Bhopal plant as well as available alternative techniques.

Table 1   Technique used in Bhopal, and alternative methods

Ref.	Bhopal plant	Available alternative
6, 9	Using MIC in the production of pesticides	Two-step process without MIC (patented by UCC)
6, 9	Storing large amounts of MIC for long periods	Storing small quantities for short periods (as in Virginia)
6, 9	Storing MIC in large tanks	Storing in small vessels (as in Texas, France, West Germany, Japan and Britain)
9, 12	Manual monitoring of important instruments	A fully computerised 4-stage alarm system (as in Virginia)

Continues...

...Continued

Ref.	Bhopal plant	Available alternative
9	No chemical inventory monitoring	A chemical inventory monitoring (as in Virginia)
9	No "knock down" tank	A "knock down" tank to take out MIC discharge to the flare tower (as in Virginia)
8, 9	Vent Gas Scrubber with limited capacity	An extra Emergency Gas Scrubber with extra capacity (as in Virginia)
8, 9	Flare tower with limited capacity	Flare tower designed for "worst possible" scenario
15	Possible to shut down VGS and flare when MIC production stopped	Continuous operation of VGS and flare
9	MIC tank kept at 0° C	MIC tank kept at -10° C (as in Virginia)
9	Possible to move or switch off refrigeration system	An extra emergency refrigeration system
6, 12	Water spray designed to reach 15 m (the flare tower was 33 m high)	A larger water spray system (as recommended after the 1982 UCC inspection)
8, 15	Carbon steel in adjacent lines	Only stainless steel (as recommended in the UC manual on MIC)
6, 9	Location close to a large densely populated area	Location downwind, outside town (as called for in the Bhopal Development Plan)
1	Alarm system directed only at the workers at the factory site	Alarm and loudspeakers pointed outwards, in the direction of the *bastis*
1, 6, 8	The alarm siren automatically stopped after ten minutes. A less noisy alarm that could not be heard outside the factory area took over. Loudspeakers were used to instruct workers	An alarm system, including loudspeakers, that reached the residents

6.3 Management Important for Safety

6.3.1 Education of workers

The education of the operators is described by Chouhan [10] and others [6, 19].

At the beginning, to be an operator in the MIC plant one had to be either a graduate in science or hold a diploma in mechanical or chemical engineering. The training of the operators at the MIC plant was later shortened from 6 months to 8 weeks. As a rule, workers and operators were given more responsibility than they had the training and competence to cope with. Vacancies at the MIC plant were filled by unskilled personnel from the closed-down parts of the plant.

In 1982, most of the original MIC operators had resigned from UCIL, because of the staffing policy of the company (see 6.3.2). After the shutdown of the naphthol plant, these workers were asked to take MIC plant training without any official letter being given. The operators' opinion on the two months of classroom training was that it was just a formality. After only 14 days of training in the MIC unit, they were asked to take independent charge of a regular plant operator's position.

The secrecy issue hampers the acquisition of knowledge by the workers [9]. Employees were never permitted, even during training, to take the company's specialised literature and safety manuals outside its premises. The manuals themselves were kept in the safe custody of the manager. Furthermore, the plant-operating manual was available only in English [7].

6.3.2 Staffing policy

Chouhan describes how in 1975, a group of new technicians were employed for an 18-month training period [10]. During this period, they were treated as casual workers. After the training, they were only placed on an hourly rate. Among the workers, it was known that the MIC plant was the most dangerous. When Chouhan accepted a job there, he got a paper about receiving six months' training. After five weeks, he was asked to stop the training and to take charge as a full-fledged plant operator. After opposing this, he got another three weeks' training.

In the matter of promotions, individuals with little experience but with unquestioning loyalty to the bosses were invariably selected before others [10]. A demand for extra safety precautions led to warnings that appointments could be terminated [10, 19].

Contract workers without safety equipment did dangerous work that should have been done by machines. Workers and operators were routinely exposed to toxic chemicals like MIC, carbon tetrachloride, trimethylamine, alpha-napthol and carbaryl dust. They seldom had the equipment recommended in the manuals [16, 20].

In 1983 and 1984, there were personnel reductions in order to cut costs. A number of workers were encouraged to take early retirement, some 300 temporary workers were laid off, and another 150 permanent workers were put in a pool to be assigned to jobs as needed [8]. The operating shifts were cut from twelve to six and the maintenance shifts from six to two [6, 8, 10]. The positions of second- and third-shift maintenance supervisor had been eliminated just a few days before the disaster [8, 21]. Employees were often assigned to jobs for which they were not qualified. If they refused, their salaries were reduced [6]. On the night of the disaster, there were no trained engineers on the site [10]. The production supervisor who was on duty had been transferred from a Carbide battery plant only a month earlier [8].

Operators were examined by the plant doctor every six months, which included an examination of blood and urine. The employees were never told the results of these examinations [10]. The UCIL management advised the workers to develop resistance against toxic substances by drinking six or seven glasses of milk a day and eating a high-protein diet of fish and eggs [13].

The personnel management policy led to an exodus of skilled personnel to better and safer jobs. The deterioration of the "beautiful plant" contributed to this.

6.3.3 Information

Most of the workers at the UCIL Bhopal plant had not received training or information about the hazards of the toxic chemicals in the plant. Residents of the adjacent *bastis* thought the plant was making "medicine" for crops. City and state authorities were not provided with information on the chemicals in the plant [6]. Although UCIL had provided Hamidia Hospital with modern resuscitation equipment, the medical staff engaged by Carbide did not get any specific training in the pathology of gas related accidents [1].

Neither UCIL nor government had prepared an evacuation plan, an emergency response system or a medical plan [17].

Company policy forbade employees to speak for the company without authorisation, especially in emergency situations [17].

6.3.4 Management of the plant

Chouhan [10] and other workers [6, 19] describe how operational problems in different parts of the Bhopal plant were solved through design modifications. Most of the modifications of the original design consisted of changes from automatic and continuous to manual and batch processes. This led to more hazardous working conditions, poor recovery of solvents, and leakage of chemicals, leading in turn to atmospheric pollution and inadequate control over the whole process of production.

In 1978, there was only one fire truck at the factory. When this was out of order, there was no fire truck at all [10].

In 1983, there was great pressure from the Danbury head office in the USA to cut expenses. Decisions were made to prolong the time between certain checks from six to 12 months, and to replace damaged stainless steel pipes with ordinary steel pipes [1]. Items that should have been replaced every six months were used for more than two years [9, 12]. Faulty instruments were not replaced. In late 1983, the principal safety systems were shut down, as the plant was not operating [1].

The plant had been shut down for a maintenance overhaul and to reduce inventories for over a month prior to the accident. The maintenance operation was almost complete and the plant was ready to resume operations by early December [8].

As the parent company held sole responsibility for all design decisions relating to the Bhopal plant, the Indian subsidiary must have secured the parent company's approval for the process modification. The former safety officer of the plant said in an interview that any design change or change in the material of construction had to be approved by the parent company in the US. According to the CBI inquiry, the approval to install a jumper line was given in May 1984 [8].

In Table 2, the management of the different safety systems of the MIC plant is described. This can be compared to the recommendations in the UC manuals on MIC [16, 20, 22].

Table 2   Safety systems and management

Safety system	Ref.	Comments
Filling of tanks	1, 6, 23	• Tank 610 was filled well above the recommended level (75–87% instead of 60%).
	1, 6, 23	• Tank 619 was not kept empty for emergency needs.
Pressure	6	• Tank 619 was not kept under pressure.
	1, 8, 23, 24	• Tank 610 could not be pressurised on October 21, November 26 or 30 and December 1.
	8	• The level indicator (LI) was faulty.
Temperature	21, 23	• No tank temperatures were logged for a long time. There was no column for it in the log book.
	8, 10	• The rule on temperature was rewritten so as to shut down the MIC refrigeration unit when production was not in process.

Continues...

...Continued

Safety system	Ref.	Comments
	10	• The temperature of MIC was not recorded on the log sheet.
	1, 8	• The alarm that should have gone off in case of any abnormal rise in temperature (15° C) in the tanks had been disconnected years earlier.
	6, 8, 23	• The refrigeration unit had been shut down one year earlier and moved and/or the freon had been drained off.
Vent Gas Scrubber	8, 10	• Safety specifications were rewritten to allow the VGS to be shut off when the plant was not in operation.
	10	• The caustic soda feed to the VGS was modified into a batch process, from the continuous process originally designed.
	6, 10, 23	• The VGS had been moved from an operating mode to a standby mode on October 23, 1984, after the MIC unit was shut down with a total MIC inventory of 83 tonnes in tanks 610 and 611. The return to an operating mode was dependent upon the operator being alert to any problem and taking prompt action to activate the circulating pump.
Lines	8	• As a section of the PVH was under repair, the jumper line had been left open.
	8	• Components of the PVH and RVVH pipelines were made of carbon steel.

Continues...

...Continued

Safety system	Ref.	Comments
Flare	6	• The backup, set to ensure that the pilot light stayed on, was discontinued to save money.
	6, 10, 23	• On November 25, a section of corroded pipe had been removed for maintenance work. A replacement pipe could have been prepared in the plant, and should have taken only four hours to install.
	10	• The pilot flame in the flare tower was lighted only when the carbon monoxide plant was running.
Analyses	10	• The rules on parameter reading were rewritten to do a reading every eight hours, as compared to every hour in 1979–80.
	23	• No analysis of caustic concentration had been made since October 23.
	23	• There was no record of analyses of MIC in tank 610 after October 19.
Alarm	8, 10, 19, 24	• The siren had been delinked from the alarm so that the operators could warn just their workmen, without "unnecessarily" alarming the people outside the plant.
	1	• The windsock, that showed the wind direction, could be seen only within the factory area. There was no windsock for the residents.
Meters	6, 10, 23	• The temperature indicator alarm was malfunctioning from just

Continues...

...Continued

Safety system	Ref.	Comments
		after the time the plant began operation, and recording of temperatures could not be done.
	6, 10	• The level indicator and alarm of the storage tank were often out of order. This created the possibility of overflow while the tank was being filled, and made it impossible to see if the level as reduced because of leakage.
	10	• Air supply to critical monitoring instruments and gauges was cut off during plant shutdowns.
Design	10	• Faulty plant design was worsened by on-site management decisions to bypass approved operating procedures to save time and money.
	8, 10, 13	• A design modification in 1983 and/or 1984 (a jumper line between the lines to RVVH and PVH) provided a route for the water to enter the MIC tank 610.
Maintenance	10, 23	• Lines were corroded
	6, 10	• Malfunctioning valves were not replaced.
	6, 10	• Faulty gauges were not repaired.
	10	• Key instruments were absent.
	1	• The stopcocks controlling access to the decontamination tower were turned off because the factory was not in service.

Continues...

...Continued

Safety system	Ref.	Comments
	10	• Since one month, the pressure control valve on tank 610 had been malfunctioning, thus allowing water to enter the tank when the connecting lines of the RVVH downstream were being washed.
Personnel	10	• Reduction of the work force and a high turnover of trained workers led to the workers not having the requisite education.
	6, 8	• The post of maintenance supervisor was eliminated.
	6, 8	• The production supervisor in charge had only recently been transferred here.
	6	• All signs and operating procedures were written in English, although many workers spoke only Hindi.
	6, 10	• Not all operators were provided with safety equipment.
	6, 10	• Maintenance work was done at night-time.
	13	• The workers were advised to "develop resistance against toxic substances by drinking six or seven glasses of milk a day and eating a high protein diet of fish and eggs".

In the weeks prior to the incident, the MIC manufacturing unit had been shut down [25]. The Sevin unit was operating using the MIC that had been stored in Tank 611.

The production of MIC was stopped on October 22. At that time, tank 610 contained approximately 42 tonnes of MIC. Tank 622 also contained an equal quantity of MIC [15]. During the period October 22 to November 30, MIC was being transferred from tank 611 to the Sevin unit, whenever required.

6.3.5 Previous warnings

In 1974, residents found a well had been contaminated. Cattle belonging to residents of Chola strayed into the area of a pool fed by a rubber pipe issuing from the factory. They drank the water and died soon after. Analyses of soil showed contamination with heavy metals. Toxic chemical substances were found in water from wells outside the plant area. UCIL did not divulge these findings [1].

In 1976, the two trade unions reacted to the pollution within the plant by sending letters to the managers of the plant and the factory inspector as well as to the Ministry of Labour of Madhya Pradesh [10]. They did not receive any answer.

In 1978, there was a big fire in the factory [10]. It showed that raw materials were being stored in places other than those designated for the purpose. It was suspected that the management had deliberately started the fire, in order to circumvent the import restrictions of alpha-naphthol laid down by the government. A report on the incident was never filed.

In 1981, a worker was splashed with phosgene. In panic, he ripped off his mask, thus inhaling a large amount of phosgene gas. He died after 72 hours. The managers blamed the worker for removing his mask. The workers' union pointed out that it was the malfunctioning valve that had led to the accident and that the worker had not been provided with a PVC overall [10]. Two others were seriously injured in this incident [17].

In January 1982, there was another phosgene leak, when 24 workers were exposed and had to be admitted to hospital. None of the workers had been ordered to wear protective masks. After this accident, the workers agitated for safer working conditions [1, 10, 17].

In February 1982, an MIC leak affected 18 workers [6].

In August 1982, a chemical engineer came into contact with liquid MIC, resulting in burns to over 30 percent of his body [6].

In October 1982, there was a leak of MIC, methylcarbaryl chloride, chloroform and hydrochloric acid [1, 6, 12]. As an operator was opening a valve in an MIC pipeline, the joint linking it to several other pipes unexpectedly broke. Evacuation of the plant was ordered. The gas plume moved toward the nearby settlement and people ran away in panic. In attempting to stop the leak, the MIC supervisor suffered intensive chemical burns and two other workers were severely exposed to the gases.

During 1983 and 1984, leaks of the following substances took place "with frightening regularity" in the MIC plant [10]: MIC, chlorine, monomethylamine, phosgene, and carbon tetra-chloride, sometimes in combination.

After the leak in 1982, the trade union printed 6,000 posters with warning texts that were distributed throughout the community [1, 6]. The Hindu union leader went on a hunger strike at the entrance to the factory. The result was that all political and trade union meetings inside the factory were banned. One UC staff member burnt the principal union's tent. In the ensuing scuffle, several people were injured. The trade union leaders were laid off. Meetings and processions were held throughout the city. As the UC staff regarded the plant as "one of the safest ships in the modern industrial fleet", the demonstrations were considered to be a campaign by agitators wanting higher salaries and shorter working hours [1].

A journalist was a neighbour of the worker who died in 1981. He had listened to the workers' discussions about the dangers at the factory – toxic gases, deadly leaks and the likelihood of explosions. After having done some research, he started to write articles in the local press, warning of the hazards associated with the plant [1, 17]. His final article, which appeared just five months before the disaster, was titled "Bhopal on the Brink of a Disaster". No one took any notice. He also sent letters where he summarised the findings of his investigations to the Chief Minister and to the Chief Justice of the Supreme Court, and requested them to close down the factory. He got no answer.

One engineer was anxious that the pipelines might crack, allowing MIC to escape and hit passing trains and their passengers. He collected information about the meteorological conditions for Bhopal from the national meteorological observatory in Nagpur, and sent it to UC in South Charleston. After computer simulation he got the answer that such a cloud would pass over the trains – which meant it would hit the *bastis* [1].

On October 21, 1984, the nitrogen pressure in tank 610 dropped to one-fifth of its normal level and none of the excess MIC could be extracted. To continue Sevin production, managers switched to tank 611. They did not investigate the cause of the pressure loss in tank 610 [1, 21]. On November 30 (or November 26 [8]3]), nitrogen pressure failed in tank 611 as well, prompting attempts to repressurise tank 610. A defective valve attached to tank 611 was repaired and tank 610 was again abandoned. Operators later told journalists that every time nitrogen was pumped in, it leaked out again through an unknown route.

6.3.6 Safety audits

Safety audits were done every year in the US and European UCC plants, but only every two years in other parts of the world [13].

Before a "Business Confidential" safety audit by UCC in May 1982, the senior officials of the corporation were well aware of "a total of 61 hazards, 30 of them major and 11 minor in the dangerous phosgene/methyl isocyanate units" in Bhopal [24].

In this audit, it was indicated that worker performance was below American standards [6]. Ten major concerns were listed; many of those are found in Table 2. It expressed alarm at the poor state and inappropriate placement of safety equipment, and at the lack of periodic checks to see that the instruments and alarm systems were functioning correctly. It also expressed concern at the alarming turnover of inadequately trained staff, unsatisfactory instruction methods and a lack of rigour in maintenance reports. Installation of an automatic sprinkler system in the MIC production zone was recommended. Three lines in the 52 pages pointed to a particularly serious mistake: a section of pipe work had been cleaned without the person in charge of the process taking the precaution of blocking off the two extremities of the pipe with special discs designed to prevent the rinsing water from seeping into other parts of the installation [1].

However, the report ended "No situations involving imminent danger or requiring immediate correction were noted during the course of the survey". UCIL prepared an action plan, but UCC never sent a follow-up team to Bhopal. Many of the items cited in the 1982 report were temporarily fixed, but by 1984, conditions had again deteriorated [6].

In mid-1984, one of the engineers, who had built the Bhopal plant, visited Bhopal again. What he saw at the plant worried him, and he tried to convey this to his superiors. They did not listen [1].

In September 1984, an internal UCC report on the Virginia plant in the USA revealed a number of defects and malfunctions. It warned that "a runaway reaction could occur in the MIC unit storage tanks, and that the planned response would not be timely or effective enough to prevent catastrophic failure of the tanks". This report was never forwarded to the Bhopal plant, although the main design was the same [1].

6.4 The State of the Safety Systems on December 3, 1984

UCC admitted in their own investigation report [23] that most of the safety systems were not functioning on the night of the December 3, 1984:
- Tank temperatures were not logged;
- The vent gas scrubber (VGS) was not in use;
- The cooling system was not in use;
- A slip bind was not used when the pipes were washed;
- The concentration of chloroform in Tank 610 was too high;
- The tank was not pressurised;
- Iron was present because of corrosion;
- The tank's high-temperature alarm was not functioning;
- Tank 619 (the evacuation tank) was not empty.

In addition, other faults are recorded:
- The meters monitoring tank E610 were showing abnormally low pressure. The reason could have been either a faulty meter or an inability of the tank to maintain pressure [9].
- The line connecting the VGS to the flare tower was master carded [9, 10].
- Many valves, vent lines, feed lines etc. were in poor condition [9, 10].

After the leakage, there were findings that give rise to several questions concerning the maintenance of the plant [9, 19].

• Water was discovered by the workers after the accident when they drained the vent lines connecting the tank E610 and the relief valve vent header (RVVH).

• Caustic soda was found in the RVVH when it was opened on December 10.

Night work has been shown to be a risk factor in accidents [26, 27]. This might have contributed to the mistakes made by the staff during the course of the leakage.

6.5 Comments

Kumar [28] points out how the culture of the British times was alive even after freedom. The elite in the small kingdoms had used the modern state organisation to retain power. Every small progress was looked upon as a heroic achievement, whereas every failure was a shame, something that had to be buried and forgotten. The elite in Bhopal tried to do the same – to bury and forget.

The deficiencies in the Bhopal plant design can be summarised as:

• A dangerous method of manufacturing pesticides chosen;

• Large-scale storage of MIC prior to selling;

• Location close to a densely populated area;

• Under-dimensioning of the safety features;

• Dependence on manual operations.

The deficiencies in the management of UCIL can be summarised as:

• Lack of skilled operators because of the staffing policy;

• Reduction of safety management because of reducing the staff;

• Insufficient maintenance of the plant;

• Lack of emergency response plans.

It was UCC that chose the design. UCC was well represented on the UCIL board. Thus, UCC can hardly avoid responsibility for the safety status of the plant.

It seems as though the management's views on safety differed from those of the workers. The chief medical officer at Bhopal said, "The safety precautions we took were the best possible. We did everything the Americans advised. In fact, we used to think that we were overdoing the safety" [13].

6.6 References

1. Lapierre, D. and J. Moro, *It Was Five Past Midnight in Bhopal.* 1st Indian ed. 2001, New Delhi: Full Circle Publishing.
2. Mac Sheoin, T., *Report on Union Carbide Corporation.* 2002, The Other Media: New Delhi.
3. MacSheoin, T., *Asphyxiating Asia.* 2003, Mapusa, Goa: Other India Press.
4. *Contempt of People. Ramifications of the Bhopal Gas Leak Disaster Case.* 1992, Delhi Science Forum: New Delhi.
5. Weir, D., *The Bhopal Syndrome. Pesticides, Environment and Health.* 1988, London: Earthscan Publications Limited.
6. *The Trade Union Report on Bhopal.* 1985, ICFTU-ICEF: Geneva, Switzerland.
7. Morehouse, W., *The Ethics of Industrial Disaster in a Transnational World: The Elusive Quest for Justice and Accountability in Bhopal.*
8. Morehouse, W. and A. Subramaniam, *The Bhopal Tragedy. What really happened and what it means for American workers and communities at risk.* 1986, New York: The Council on International and Public Affairs.
9. *Bhopal Gas Tragedy.* 1985, Delhi Science Forum: New Delhi.
10. Chouhan, T.R., *Bhopal: The Inside Story. Carbide workers speak out on the world's worst industrial disaster.* 1994, New York: The Apex Press.
11. Ramachandran, C.R., *Immediate Post Industrial Disaster Management,* in *Refresher Course for Top Executives. Management of Chemical Accidents.* 1994, Disaster Management Institute: Bhopal.
12. Chauhan, P., S., *Bhopal Tragedy. Socio-legal Implications.* 1996, Rawat Publications: Jaipur.
13. Jones, T., *Corporate Killing. Bhopals Will Happen.* 1988, London: Free Association Books.
14. *Carbide's war gas tests right under DST's nose!,* in *Bhopal: Industrial Genocide?* 1985, Arena Press: Hong Kong.

15. Varadarajan, S.E.A., *Report on Scientific Studies on the Factors Related to Bhopal Toxic Gas Leakage*. 1985, Indian Council for Scientific and Industrial Research: New Delhi.

16. *Carbon monoxide, Phosgene and Methyl isocyanate. Unit Safety Procedures Manual*. 1978, Union Carbide India Limited, Agricultural Products Division: Bhopal.

17. Cassels, J., *The Uncertain Promise of Law: Lessons from Bhopal*. 1993, Toronto: University of Toronto Press Inc.

18. Morehouse, W., *Unfinished Business. Bhopal ten years after*. The Ecologist, 1994. **24(5)**.

19. *Other Workers Speak Out: Testimonies from Union Carbide Bhopal Plant Personnel*, in *Bhopal et al. The Inside Story*, T.R. Chouhan, Editor. 1994, The Apex Press: New York.

20. *Methyl Isocyanate. Union Carbide F-41443A - 7/76*. 1976, Union Carbide Corporation: New York.

21. *The Machine Stops: Runaway Reaction*. 2000, Jan 5, http://home.earthlink.net/~wroush/disasters/bhopal3.html.

22. Behl, V.K., et al., *Operating Manual Part-II. Methyl Isocyanate Unit*. 1979, Union Carbide India Limited, Agricultural Products Division: Bhopal.

23. *Bhopal Methyl Isocyanate Incident. Investigation Team Report*. 1985, Union Carbide Corporation: Danbury, CT.

24. *The Bhopal Gas Tragedy 1984– ? A Report from the Sambhavna Trust, Bhopal, India*. 1998, Bhopal People's Health and Documentation Clinic: Bhopal.

25. Kalelkar, A.S. and A.D. Little. *Investigation of Large-magnitude Incidents: Bhopal as a Case Study*, in *The Institution of Chemical Engineers Conference on Preventing Major Chemical Accidents*. 1988. London.

26. Mitler, M.M., et. al., *Catastrophes, Sleep and Public Policy: Consensus report*. Sleep, 1988. **11**: p. 100–9.

27. Akerstedt, T., *Increased Risk for Accidents at Night Work*. Lakartidningen, 1995. **92**: p. 2103–4 [In Swedish].

28. Kumar, K., *Princes and Pesticides*. South Asia - Political and cultural magazine, Sweden, 1985. **3**(7–8) [In Swedish].

THE EVENT PHASE: THE LEAKAGE AND ITS IMMEDIATE EFFECTS

7.1 The Leakage

7.1.1 The course of the leakage

The first UC team report [1] was published in March 1985. In Kalelkar's report [2], written four years after the leakage, the log records have been carefully examined and interviews with management and workers analysed. One chapter deals with the contradictions in the statements.

The CSIR report [3] was formally released around 15 years after the leakage.

Other reports on the course of the leakage have been written by ICFTU-ICEF [4], Delhi Science Forum [5], Kulling and Lorin [6], and Chouhan [7, 8]. Sometimes, these reports are contradictory. Table 3 is a compilation of statements from the seven reports and other material.

Table 3   The course of the leakage

Time	Ref.	Event
20.30	7	• Water washing of lines began. A slip bind was not used.

Continues...

...Continued

Time	Ref.	Event
	9	• Water washing of lines began. Slip-binds were not used. When the water did not come out through the drain-cocks, the water was cut off and the filters cleaned. Then the water was released again.
21.15	10	• An operator was asked to flush the lines. He opened a nozzle on one of the pipes and inserted a water hose.
21.30	4	• Water washing of lines began. A slip bind was not used. When the operator noticed that no water was coming out of the bleeder lines, he shut off the flow, but the MIC Plant supervisor ordered him to resume.
	11	• An operator started pumping water under high pressure into four lines downstream of the MIC storage area, all of which were connected to the RVVH. According to standard operating procedures, the maintenance crew had prepared the job (i.e. closed the isolation valve between these branches and the RVVH). The outflows of bleeder valves were not releasing water at the same rate at which it was being pumped in. Two valves were completely clogged and the others partly clear. He stopped washing and reported the problem to the supervisor.
	10	• The operator turned on the water. As no water was coming out the overflow line, he turned off the hose. The supervisor ordere him to resume.
21.45	11	• The operator resumed washing the lines. But the bleeders remained obstructed and soon water accumulated in the pipes.
22.00	7	• Approximate time water entered Tank 610. Reaction begins.
22.15	2	• Final transfer of MIC from Tank 611 to the SEVIN unit according to one log, that seems to have been changed.

Continues...

...Continued

Time	Ref.	Event
22.20	1, 6, 7	• Pressure of tank 610 noted as 2 psi.
	6	• An operator was told to use water to clean lines close to the MIC tank.
22.30	9	• The operator was told to keep the water running, and that the night shift would turn the tap off.
	10	• The new operators came on duty. They noted and logged in the pressure of tank 610 as 2 psi. Temperature was not recorded.
22.45	2, 4	• Shift change. The unit was shut down for around half an hour.
		• The third shift began to suffer throat and eye irritation from a MIC leak close to the area where the lines were being washed.
23.00	9	• Shift change.
	7	• MIC leak first reported by field operator in area near the vent gas scrubber.
	1, 6	• Pressure of Tank 610 noted as 10 psi.
	1	• The operator said later that the pressure was 2 psi.
	9	• The operator noticed that the recorded pressure 2 psi was three hours old.
23:30	2	• Last transfer of MIC to the SEVIN unit. It seems to have come from tank 610 instead of tank 611.
	2	• The gases came out of the VGS.
	7, 8	• The operators noticed water leaking along with MIC in the MIC process area.
	3	• The operators on ground level noticed dirty water spilling from a higher level in the MIC structure and MIC in the atmosphere. The MIC and dirty water were coming out of a branch of the RVVH. The pressure safety valve had been removed and the open end of the RVVH branch line was not blinded.
	2, 8	• The operators reported the leak to the MIC supervisor and began to search for it in the

Continues...

...Continued

Time	Ref.	Event
		MIC structure. The leak was considered "normal".
24.00	9	• The operators found brownish water and steam coming out from a drain-cock eight yards off the ground. The supervisor recommended turning off the water taps after the tea break. The team left for the staff cafeteria.
	7, 8	• The operators found a section of open piping located on the second level of the structure near the VGS. They fixed a fire hose so that it would spray in that direction and returned to the MIC control room believing that they had successfully contained the MIC leak.
	3	• The operators went to the control room and informed the plant superintendent and the supervisor that there was a MIC leak. They were advised to spray water around the point of leakage.
00.15	2	• The tea break began. According to some, the alarm signalling the major release went off for only several minutes. Others stated that the tea period in the control room was normal. When a tea boy entered the MIC control room, he noticed that the atmosphere was tense and quiet. The operators refused the tea.
	2	• The two supervisors and the superintendent were, against the rules, taking a break together in the plant's main canteen for 45 minutes prior to the release. Here, they received word of the incident.
	2	• The transfer from tank 611 to SEVIN unit. The UC investigation team concluded that this was an attempt by the MIC operators to remove water from the tank.
	2	• It is clear that the MIC operators knew at least 30-45 minutes before the release that

Continues...

...Continued

Time	Ref.	Event
		something was seriously wrong, and that several had taken action in an attempt to forestall the problem.
	1, 6, 7	• The field operator reports the continued release of both water and MIC in the MIC process area. Water was sprayed on the leaking point. The tank pressure reading was noted as 30 psi and rapidly rising. Within moments, the pressure reading exceeded 55 psi, the top of the scale.
	3, 9	• The control room operator observed on the Pressure Indicator (PI) that the pressure was shooting up and was in the range 25–30 psi/g.
	3, 9	• Between 00.15 and 00.30 hours, PIN showed a reading beyond the maximum of the scale, i.e., higher than 55 psi/g.
	6	• Values up to 100° C and 100 psi have been mentioned.
	3	• The control room operator went to the storage area and heard a hissing sound from the safety relief valve (SRV), indicating that the SRV had popped off. He noticed that the local temperature and pressure transmitters were indicating values beyond their ranges (i.e., +25° C and 55 psi/g).
	9	• Two operators went to the tank 610. The pressure gauge indicated 55 psi. Some movements were felt inside the tank. The smell of MIC, phosgene and MMA was noticed. A geyser burst from the spot where the gas leak was detected. The operator set off the general alarm siren. The supervisor left his tea and rushed to the tank. The tank and the concrete were trembling, cracking and creaking.
	1, 3, 6	• The control room operator called his supervisor and ran outside to the tank. He

Continues...

...Continued

Time	Ref.	Event
		heard rumbling sounds from tank 610, a screeching noise from the safety valve and felt heat radiating. As he ran back to the control room, he heard the cracking of the concrete over the tank. As soon as he returned to the control room, he turned the switch to activate the VGS The flow meter did not indicate that caustic circulation had been started. The operator did not go into the unit to check the pump and verify whether there was a flow.
	3	• A gaseous cloud was seen to be coming out from the stack by the field operator.
	7, 8	• Water washing of lines was stopped. All the water from the lines came out through the open bleeders. Near these open bleeders, MIC was detected. The alarm glass was broken to start the loud factory siren. After a few minutes, the loud siren was turned into a muted siren.
00.20	1, 6	• The MIC Production Supervisor notified the Plant Superintendent, who was in the formulation area, of the release.
00.25	1, 6	• The Plant Superintendent arrived in the MIC unit and found a lot of MIC in the atmosphere.
?	9	• All communication between tank 610 and tank 611 was shut off by the supervisor and the operator. Tank 610 stood vertically, fell and stood up again, but did not burst. A second geyser erupted from a ruptured pipe at ground level.
00.30	7	• Tank 610 showed a noticeable heat increase and began to make a rumbling sound. The concrete casing of the tank then split due to the expansion of the tank walls caused by the increase in pressure. The rupture disc

Continues...

...Continued

Time	Ref.	Event
		broke, the safety valves for tank 610 popped and the bulk of the tank contents was released through the vent gas scrubber.
	11	• The estimated time of the leak.
	9	• The water was cut off.
	3	• The siren was sounded and the plant was alerted to the leakage.
?	9	• The siren was turned off and the workers alerted through the loudspeakers.
	11	• The workmen were instructed to evacuate the plant. As the windsock indicated that the wind was blowing westward, they were instructed to move east.
00.45	1	• The Supervisors' Log records that Derivatives Unit operations were suspended because of the high concentration of MIC in the area.
00.50	8	• The alarm began inside factory, alerting the workers to a hazardous leak. The emergency squad tried to control the leak by massive water spraying, but the water did not reach the site for the leak was not high enough.
	11	• A workman sounded the toxic gas alarm. In accordance with the plant's emergency procedures, the control room operator immediately switched off the siren and made an announcement over the plant's system informing the workmen. He then restarted only the internal alarm.
01.00	12	• A worker broke the alarm glass and started the factory siren.
	6, 8	• When the plant superintendent came back from his smoking break, he ordered that the loud siren be stopped.
	1	• A derivative unit operator turned on the Toxic Gas Alarm.

Continues...

...Continued

Time	Ref.	Event
	1, 3, 6	• The Plant Superintendent and the MIC operator verified that MIC from Tank 610 was being emitted from the VGS stack to the atmosphere. They turned on the fixed fire-water monitors and directed them to the stack. Water streams were also directed on the MIC tank mound and on the relief valve line to the VGS for cooling. Steam came from the cracks in the concrete indicating the MIC tank was hot.
	3, 6, 9	• The water pressure was not high enough for the water to reach the point of emission.
	6	• One of the workers tried to climb up the construction to close the leakage. He was so exposed to the gases that he fell down and fractured several bones.
	2	• Within 15 minutes of the major release, the MIC supervisor called the MIC production manager at home and told him that water had got into an MIC tank.
01.30	8	• Workers began to flee the factory premises to save their lives.
	1, 6	• Sometime between 1.30 and 2.30, the safety valve restated, indicating a tank pressure below 40 psi/g, and the emission of MIC stopped.
02.00	8	• The workers realised that the toxic release was affecting the communities outside the plant. They insisted that the plant superintendent should restart the loud siren, which he finally did.
	11	• Most of the contents of tank 610 had escaped.
02.15	7	• The gas leak stopped.
03.00	3	• The SRV of tank 610 is reported to have sat back and the gas also stopped coming out from the stack.

Continues...

...Continued

Time	Ref.	Event
	8	• People from outside came to the factory dispensary for treatment.
05.30	1, 6	• Tank 610 was hot to the touch (45–60° C).
06.00	1, 6	• The thermometer on the VGS caustic accumulator read 60° C, indicating that an MIC reaction had taken place.
07.00	8	• Both pressure gauges on the tank were twisted out of range. The tank manhole was very hot.
Morning	3	• One witness noticed that a pressure indicator on Tank 610 was missing and that no plug had been inserted in the opening. A water hose was lying nearby. Other witnesses contradicted these statements.
After leakage	5, 8	• Water was discovered by the workers after the accident when they drained the vent lines connecting the tank 610 and the relief valve vent header (RVVH).
Dec 10th	5, 8	• Caustic soda was found in the RVVH when it was opened.

7.1.2 The chemical reaction

In the CSIR report [3] the reaction in the tank is discussed. Phosgene is always present in MIC, as an inhibitor of polymerisation. When water entered Tank 610, it would have reacted with phosgene and methylcarbamyl chloride (MCC) to produce hydrochloric acid (HCl). For this, relatively small amounts of water, even a few litres, would be enough to reduce the phosgene to very low levels. HCl would react with metal particles to produce ionisable metal chlorides. HCl and the metal chlorides would, in the absence of phosgene, catalyse into a violent and explosive polymerisation. The heat would promote a chain reaction, leading to a very rapid increase in temperature, vaporisation, increase in pressure and leakage of gas.

The UC team [13] stresses that the corrosion rate would have increased markedly when the temperature increased, because of the presence of an abnormally high level of chloroform. Thus, more iron would be produced and catalyse an exothermic trimerisation of MIC. The violence of the reaction would increase further.

The amount of water entering the tank is disputed. One estimate is that at least 200 litres (kg) of water entered the tank. The CSIR says that the chemical analysis of the tank residue clearly shows the evidence of the entry of approximately 500 kg of water. The UC team talks about 120 to 250 gallons of water (480–1,000 kg) [13].

The temperature in the tank is a very important question, as MIC decomposes to hydrogen cyanide at higher temperatures (see 7.2.3). UCC has maintained that the temperature would have been 0–250° C, while others claim it would have been 400 or even 540° C [11, 13]. The CSIR report [3] points out that information from the mechanical examination of the tank indicates that the pressures may have reached 11 to 13 kg/cm^2 with the corresponding temperatures in the range of 200 to 350° C. From the products found in the residue, the calculated amount of heat of chemical reactions and the extent of bulging of the exhumed tank, it is surmised that the temperature in the tank rose above 250° C.

In the UC report [1], the tank temperatures are not described and hydrogen cyanide is not discussed.

The CSIR team found that the reactions of MIC with small quantities of water and chloroform at 250° C give all the products found in the solid residues in Tank 610, except tetramethyl biuret [3].

Various Indian investigative journalists argue that there were, in fact, two runaway reactions in Tank 610 [11]. The first was the reaction of MIC with itself, catalysed by iron contaminants washed into the tank. The second reaction was of MIC with water. The UC manual indicates that a runaway reaction of MIC with water by itself occurs only after 23 hours at 20° C. But such a reaction is greatly accelerated and can take place in just a few hours when MIC is reacting with itself, catalysed by iron.

7.1.3 The direct cause—four theories

7.1.3.1 The water washing theory

This is the workers' theory, supported in the Government's official report [3] and by the chief lawyer of UCIL in the Bhopal court [7]. It is described in detail in the Trade Union Report [4], by Morehouse and Subramaniam [11], and by Chouhan [7, 8]. It is discussed by Cassels [14], Jones [13] and in the UC team report [1]. Lapierre and Moro [9] have interviewed operators and workers.

Those in charge of the MIC plant on the evening December 2nd were not familiar with the factory's complex maintenance procedures, and they knew nothing about MIC or phosgene. The supervisor was convinced that there could not be a leak when production had been stopped.

The supervisor from the day shift had left instructions on flushing the pipes leading from the MIC tanks to the vent gas scrubber with water. He forgot to mention the slip-binds that should have been placed at each end of the pipes. When the worker placed the stopcocks, he was not sure that they tightened completely, because of corrosion and rust. The water did not come out of the drain-cocks, and he found that the filters were blocked with metal debris. He cut off the water. The supervisor told him to clean the filters. When he turned on the water, it came out through three of the four drain-cocks. He was told to keep the water running, and that the night shift would turn it off.

The workers maintain that entry of water through the plant's piping system during the washing of lines was possible because a slip-bind was not used, the downstream bleeder lines were partially clogged, many valves were leaking, and the tank was not pressurised. Carried with the water were iron rust filings from corroding pipe walls, residue of the salt compounds that had blocked the lines being washed, and other contaminants.

The water, which was not draining properly through the bleeder valves, may have built up in the pipe, rising high enough to pour back down through another series of lines into the MIC storage tank. Once water had accumulated to a height of 20 feet (6.1 m), it could drain by gravity flow back into the system. Alternatively, the water may have been routed through another standby "jumper line" that had only recently been connected to the system. Indian scientists suggested that additional water might have been introduced as a "back-flow" from the defectively designed vent-gas scrubber.

The UC team says that this was impossible. In order for the water to reach the MIC tank during washing, it would have had to travel through dozens of metres of piping, pass through several valves, and finally climb 3.5 metres to reach the tank opening. The UC team also says that "entry of water into tank 610 from this washing would have required simultaneous leaks through several reportedly closed valves, which is highly improbable". Chouhan points out that "grossly inadequate maintenance would permit water to pass even through closed valves because of malfunctioning".

The CBI's officials, who disconnected the pipeline from the storage tank on the morning after the leak, drained out as much as 27 litres of water from the structures on 14 February [13]. Water was found in almost all the connecting pipelines tracing the entire route from the point of washing to the Relief Valve Vent Header. This supports the water washing theory.

UCC's team spent 24 days in India, and continued its work for months thereafter [11]. However, they were not permitted to interview UCIL employees, and they were not allowed to examine the pipelines to determine how water got into the tank. UCC's only investigative work consisted of chemical investigations, and the results have still not been released [13].

7.1.3.2 The direct entry theory

The UCC hypothesis was that the runaway reaction in the tank occurred "when a substantial amount of water was introduced". The theory is that somebody deliberately connected a hose to a pressure gauge [2, 4]. In a press conference in March 15th, 1985, it was hypothesised that "water could have been introduced inadvertently or deliberately directly into the tank through the process vent line, nitrogen line or other type of line". It is also said that someone might have connected a tube to the water line instead of the nitrogen line.

After objections from the workers, it was admitted that the company's investigative team did not find any evidence for such a connection [4]. Some journalists pointed out that the reaction would have occurred 23 hours later, and that there are no valves, vents or bleeders to which a hose could be put [11]. A senior UC official should have told a Congressional committee "that the IC tank line fittings are colour-coded and that the water line couplings are incompatible with the gas line couplings that go into the tank".

All the same, UCC blamed first a Sikh terrorist, then "a disgruntled worker". From Kalelkar's report [2], it is obvious that this worker has been identified by UCIL, but he was never reported. The worker has identified himself [8, 9], and says that the workers knew that putting water directly into the MIC tank would be extremely dangerous for the person himself. So any worker would avoid sabotage. He also says that if there was sabotage, the culprit would be the management, which was responsible for supervising the safety precautions at the MIC plant.

One UCC spokesperson could not think of a motive, as there were "no radicals or groups like that". Some US financial analysts dismissed UCC's theory. Anderson himself admitted to a US Congressional panel that he had "no evidence whatsoever that sabotage was behind the disaster" [13].

The CSIR report in December 1985 notes: "The scientific analysis shows that any addition of water alone, even deliberately, could not lead to such an accident. Anyone wishing to cause an accident of this nature would have to be presumed to have very substantial knowledge and information that metal contaminants would already be present and that the alarms and safety systems installed for containment were grossly inadequate" [3].

UCIL has later admitted that the sabotage theory was false, arguing instead that three employees caused the disaster through negligent behaviour [15].

7.1.3.3 *The economy theory*

The UCIL factory was running at a loss and it was decided that the factory should be closed down and sold. However, at that time, to close a plant in India needed the permission of the Government. It was suggested that Mr Warren Anderson wanted a minor accident, so that the Government of India would allow him to close the plant. In this scenario, the leakage appeared to have been put on the stage on purpose, but seems to have run out of control.

7.1.3.4 The warfare test theory

Jones [13] describes this theory as mainly proposed by pro-Soviet elements in India: that it was a deliberate chemical warfare experiment on the part of the USA.

A Research and Development unit was set up in Bhopal in 1976. The centre, the biggest in Asia, had five insect-rearing laboratories and a two-hectare experimental farm for testing chemical agents. Here, new molecules were synthesised and tested. It appeared that the UCIL had been conducting field studies using new chemical agents without getting the projects cleared by the top-level committee from 1975, where all collaborative research efforts should be screened from a "security" angle [5].

The closeness of chemicals for peace to chemicals for war was highlighted by Indian charges that the centre was involved in chemical warfare experimentation [13]. The Delhi Science Forum said that "much of the results of the findings in these areas are also quite likely to have never been published, given the enormous significance such data has for chemical warfare", and also that "the centre's studies covered the grey area between agricultural research and anti-crop warfare" [5].

This twist was highlighted by reports about the presence of chemical warfare experts at Bhopal studying MIC's effects [13]. For example, it is known that from the Pentagon, a medical doctor was sent to collect military intelligence regarding the effects of the leaked gases. From Sweden, two doctors were sent to make a report for the National Defence Research Institute [6].

7.1.3.5 Other theories

There are other theories about how the water got in the tank [11]. Water could have entered at some point in the nitrogen line near the tank, through the refrigeration line, or directly through the process vent system. However, serious problems are manifest with these theories.

7.1.4 Comments

The direct cause of the gas leak was the large amounts of water that entered Tank 610. A run-away reaction started, which was speeded up because of contaminants, high temperatures and other factors.

The two main theories as to how the water entered the tank are the sabotage theory and the water washing theory. UC has also pointed out contradictions in the statements from the witnesses [2].

However, sabotage would have been improbable if
• maintenance had been good;
• the safety systems had been working;
• the saboteur would have wanted to save his own life and health.

After these studies, the author is convinced that the water washing theory is the plausible theory. The actions of UCC shows the weakness of the sabotage theory.

Even if the supervisors had been properly trained and had acted earlier, they would not have been able to control the leakage, as the different safety systems were either under-dimensioned or not working. The leakage stopped when Tank 610 was empty.

A very important factor is the temperature of the tank. It is in the UCC's interest to maintain that the temperature was never over +200° C, which would mean that only very small amounts of hydrogen cyanide were formed. However, it seems very likely that the tank temperature was far above this point.

The leakage, irrespective of the initiating cause, would not have reached this magnitude if
• The MIC had been stored in several small tanks instead of two big ones;
• The proper materials had been chosen for the pipeline system, so it would not have contributed to the contaminants;

- The maintenance had been appropriate, so the risk of contaminants would have been minimised;
- The safety systems had all functioned as planned at start of the plant;
- The safety rules had all worked as planned at start of the plant;
- The stated safety precautions had been followed;
- The operators and workers had been properly educated.

7.1.5 References

1. *Bhopal Methyl Isocyanate Incident. Investigation Team Report.* 1985, Union Carbide Corporation: Danbury, CT.
2. Kalelkar, A.S. and A.D. Little. *Investigation of Large-magnitude Incidents: Bhopal as a Case Study.* In *The Institution of Chemical Engineers Conference on Preventing Major Chemical Accidents.* 1988. London.
3. Varadarajan, S. et al, *Report on Scientific Studies on the Factors Related to Bhopal Toxic Gas Leakage.* 1985, Indian Council for Scientific and Industrial Research: New Delhi.
4. *The Trade Union Report on Bhopal.* 1985, ICFTU-ICEF: Geneva, Switzerland.
5. *Bhopal Gas Tragedy.* 1985, Delhi Science Forum: New Delhi.
6. Kulling, P. and H. Lorin, *The Toxic Gas Disaster in Bhopal December 2-3, 1984.* 1987, Forsvarets Forskningsanstalt (National Defence Research Institute): Stockholm, Sweden. [In Swedish.]
7. Chouhan, T.R., *Bhopal: The Inside Story. Carbide workers speak out on the world's worst industrial disaster.* 1994, New York: The Apex Press.
8. *Other Workers Speak Out: Testimonies from Union Carbide Bhopal Plant Personnel,* in *Bhopal et al. The Inside Story,* T.R. Chouhan, Editor. 1994, The Apex Press: New York.
9. Lapierre, D. and J. Moro, *It Was Five Past Midnight in Bhopal.* 1st Indian ed. 2001, New Delhi: Full Circle Publishing.
10. *The Machine Stops: Runaway Reaction.* 2000, Jan 5, http://home.-earthlink.net/~wroush/disasters/bhopal3.html.
11. Morehouse, W. and A. Subramaniam, *The Bhopal Tragedy. What really happened and what it means for American workers and communities at risk.* 1986, New York: The Council on International and Public Affairs.
12. *The Bhopal Gas Tragedy 1984– ? A report from the Sambhavna Trust, Bhopal, India.* 1998, Bhopal People's Health and Documentation Clinic: Bhopal.

13. Jones, T., *Corporate Killing. Bhopals Will Happen.* 1988, London: Free Association Books.
14. Cassels, J., *The Uncertain Promise of Law: Lessons from Bhopal.* 1993, Toronto: University of Toronto Press Inc.
15. Morehouse, W., *The Ethics of Industrial Disaster in a Transnational World: The Elusive Quest for Justice and Accountability in Bhopal.*
14. Varadarajan, S.E.A., *Report on scientific studies on the factors related to Bhopal Toxic Gas Leakage.* 1985, Indian Council for Scientific and Industrial Research: New Delhi.
15. *Bhopal Gas Tragedy.* 1985, Delhi Science Forum: New Delhi.
16. Bidwai, P., *The Poisoned City-Diary from Bhopal,* in *Bhopal: Industrial Genocide?* 1985, Arena Press: Hong Kong.

7.2 The Contents of the Cloud and its Effects

7.2.1 Theories

The exact contents of the cloud are still not known – partly because Union Carbide has not released the information they have available. But the list of possible components is very long (appendix 2).

When asked, on the morning of December 3, by the Commissioner, the highest civil authority for Bhopal, if the gas contained cyanide, the Chief Engineer admitted, "MIC breaks down to several gases, among them hydrocyanide acid, at very high temperature". The workers who went to the tank during the leak noticed the smell of bitter almonds [1].

The doctors' first assumption was that phosgene had leaked. This was later said to be impossible, because phosgene was never stored, and the MIC plant had been shut down [2]. However, in 1981, a worker died from phosgene inhalation, although phosgene ought not to have been present in the lines after the MIC plant was shut down [3]. Also, phosgene was smelt close to the tank during the release [1].

In a paper [4] that was probably written by ICMR directly after the leakage, it is not only MIC that is described, but also phosgene, hydrogen cyanide and carbaryl (Sevin).

When cyanide was found in the blood of dead victims, there was a discussion about whether the cyanide had come from MIC in the body or from outside the body. UCC, Bayer AG which also produces MIC, and Dr Jaeger of the WHO stated that the cyanide did not come from the MIC. This means that it was not only MIC that escaped from the tank, but also other gases. This has been verified by the findings of high concentrations of hydrogen cyanide in air samples close to the tank two to three days after the leakage [3].

On the basis of investigations of post-mortem blood and tank residues [5, 6], ICMR in their 1991 report drew the following conclusions: "MIC trimer was found to be present in preserved autopsy samples of gas tragedy victims along with DMI, dione and several other unidentified compounds in the category of molecular weight of 269, 279, etc. The presence of

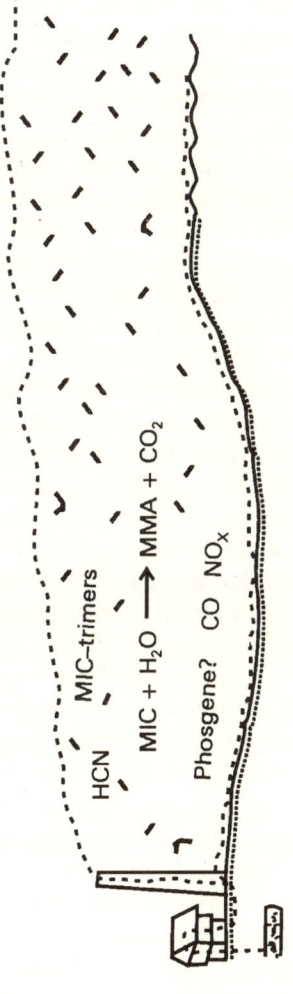

Figure 1   *The cloud that spread over Bhopal*

MIC trimer and other similar peaks of the tank residue in the blood-stream of victims indicate a definite evidence of entry of mixture of gases and particulates into the body system contrary to the statements of the manufacturers of MIC."

Union Carbide has still not released any kind of information on the composition of the cloud. In the manual [7], there is a list of compounds that are derived from burning MIC. The highest concentration would be nitrogen (N_2), the next, carbon dioxide (CO_2), and water. Smaller amounts of carbon monoxide (CO), O_2, H_2, OH-, and NO would also be released. The Material Safety Data Sheet states that MIC" can undergo a 'runaway' reaction if contaminated" and that the thermal decomposition of MIC" may produce HCN, NO_x, CO and/or CO_2," [8].

In the manual [7], a range of different reactions are described. But the names and properties of the resulting products are not described. It seems, however, likely that at least some of those products would also have been present in the cloud.

In the CSIR report [9] it is assumed that the gaseous products from an MIC-water reaction would be mainly MIC, carbon dioxide (CO_2), methyl chloride, methylene dichloride and carbon tetrachloride (see 7.1.2). Some alkylamines may have been present in very small amounts. Phosgene and methylcarbamyl chloride, present in very small amounts, are readily hydrolysed with water to give hydrogen chloride (HCl).

Subramaniam [10] also discusses the possible components of the gas cloud. In his opinion, nearest the factory, the cloud would have been mainly composed of MIC and trimers of MIC, but would also have contained hydrogen cyanide (HCN), oxides of nitrogen (NO_x), carbon dioxide (CO_2) and carbon monoxide (CO), all of which replaced the air. As phosgene is always present in the MIC tank, it is most likely that phosgene was also released.

Subramaniam is of the opinion that once the bulk of the MIC released came into contact with the moisture in the air, it would have been converted in the atmosphere into monomethylamine (MMA) and more carbon dioxide. It is likely that the heaviest (crystals and liquids) and the lightest (HCN, NO_x, CO_2 and CO) would have been confined to the area adjacent to the plant. Therefore, it is highly improbable that cyanide would have been widely dispersed. Stable compounds might have been carried away by the wind, to the lakes and the soil. As most of these compounds are heavier than air, concentrations would have been highest close to the ground.

Estimates are based on the leakage of 30 tonnes of MIC spread over the whole volume of a two kilometre radius up to a height of 30 m and they indicate that the air would have had about 0.03 ppm* of MIC [11]. As MIC is heavier than air, it is likely that its concentration would have been much higher near the land surface.

Singh and Ghosh have provided simulations of exposure concentrations at various distances downwind of the plant. A total of 27 sites were identified with ground level concentrations ranging from 85.6 ppm to 0.12 ppm with a median of 1.8 ppm [12].

Lapierre and Moro [1] point out that MIC is twice as heavy as air, and formed the base of the cloud. Phosgene, hydrocyanide acid and MMA are less dense than MIC, and could spread more rapidly and further. The cloud was not homogenous.

According to Morehouse [13], in a controlled environment, MIC would form the lowest layer, above that would be MMA, then air and HCN, and finally, the oxides of nitrogen, carbon dioxide, and carbon monoxide. But as gas escaped, both at ground level and from 100 feet, the heavier gases descending from above would trap the lighter elements in the gas released at ground level.

7.2.2 Isocyanates

The knowledge in 1984 about the chronic damage caused to human beings by isocyanates in general was based on two small but prolonged series of studies of 35 British firemen exposed to toluene diisocyanate, which was published in 1976 [13].

At the University of Lund in Sweden, research on isocyanates has been under way for more than 10 years [14]. New methods for measuring the concentration of different isocyanates in air and in the body have been developed. The results have attracted attention from the trade unions.

---

* Parts per million (concentration)

Isocyanates are a group of nitrogen compounds that are among the most common industrial chemicals. They have been used for more than 50 years. The production of isocyanates has doubled in less that 20 years, and is now around two million tonnes in Western Europe alone. Sweden has had regulations on isocyanates since 1980.

Since the early 1950s, it has been known that isocyanates can cause trouble for the eyes, nose, deeper airways and the skin. Research has shown that isocyanates can cause a range of diseases: asthma, bronchitis, pneumonia and eczema. Isocyanate asthma is the dominating professional asthma, and 5–10 percent of those who are exposed to low to moderates doses are expected to develop asthma. Among the heavily exposed, as many as 15–30 percent may develop asthma. The asthma can be chronic and exaggerated, even if exposure is ended. Antibodies to isocyanate can be found in affected as well as in non-affected persons. There are also affected persons who do not have antibodies.

Isocyanates are highly reactive and can be demonstrated to be attached to proteins in the body, thus being transformed into amines. The research is very complicated, partly because there are many more types of isocyanates than has been shown and partly because the isocyanates are transformed into different forms in the body than in the air.

The threshold limit value for isocyanates is 0.005 ppm or 0.04 mg per cubic meter air. The limit value for hydrogen cyanide is 5 mg per cubic meter air. As shown above, biological limit values ought to be used for isocyanates.

7.2.3 Methyl-isocyanate

According to reports seized from the research and development centre of the plant at Bhopal as well as documents traced from other offices of the firm, the corporation had conducted a number of experiments on animals and plants, and was well aware of the effects of MIC [3]. It is likely that they had information not only on short-term effects, but also on medium and long-term effects.

Tests on rats turned out to be so terrifying that the company banned publication of their work [1]. Experiments had shown that animals exposed to MIC vapours died almost instantaneously. The vapours destroyed the respiratory system with lightning speed, caused irreversible blindness and burnt the pigment of the skin.

Studies as early as 1963 and 1970 showed that under the influence of heat, MIC breaks down into several potentially fatal molecules, including hydrocyanide [1]. Thiosulphate would be an effective antidote. Confidential research in 1963 concluded that "methylisocyanate appears to be the most toxic member of the isocyanate family" and that it "is highly toxic by both the peroral and skin penetration routes and presents a definite hazard to life by inhalation" [13]. The 1970 research stated that MIC "is highly toxic by inhalation, an irritant to humans at very low vapour concentrations, and a potent skin sensitizer".

The properties of methyl isocyanate are described in the manuals from Union Carbide [7, 15], in the trade union report [16] and by the Swedish National Institute for Working Life [17].

MIC ($H_3C-N=C=O$) is a colourless liquid with an odour like tear-gas, slightly soluble, highly reactive when in contact with water, and lighter than water. The vapour is heavier than air. MIC is reactive, toxic, volatile and flammable. The flash-point of MIC is $-18°$ C, and a concentration of only 6 percent in air is explosive. MIC boils at $39.1°$ C. Its reactivity is inhibited by phosgene and increased by metals, acids, basic substances and amines.

The threshold limit value set by the American Conference on Government Industrial Hygienist was 0.02 ppm. MIC is defined as a poison by inhalation in DOT regulations (US Department of Transportation) [18].

At 1 ppm in the atmosphere, people's eyes will start to water and, at that stage, there may already be a high enough concentration to cause serious internal damage. When four healthy volunteers were exposed to MIC for brief intervals (up to 15 minutes), they reported some irritation, with resulting tears, at 2 ppm. As the concentration increased, symptoms worsened until they became unbearable at 21 ppm [3].

In the UC manuals from 1976, the effects on health are described [7, 15]: "MIC may cause severe or permanent injury in contact with eyes or skin. If inhaled or swallowed in sufficient quantities, death may result. MIC acts like tear-gas, but is (many times) more lethal. MIC is a poison by inhalation ... and is intensely irritating to breathe. It causes severe bronchospasm and asthma-like breathing. ... It should also be regarded as an oral and contact poison. Skin contact can cause severe burns. The liquid will seriously injure the eyes, even when it is diluted with a non-toxic liquid to a one percent concentration".

A runaway reaction is described in the UCC manuals:
- "Water reacts exothermically to produce heat and carbon dioxide. As a result, the pressure in the tank will rise rapidly if MIC is contaminated with water. This reaction may begin slowly, especially if there is no agitation, but it will become violent."
- "MIC reacts vigorously with contaminants such as water, acids, alkalis and amines and can polymerise rapidly if in contact with iron, steel, zinc, tin, galvanised iron, copper and its alloys. Stainless steel is safe."

It has been calculated that the leakage of one gallon (4.5 litres) of MIC can cause health problems two miles (3 kilometres) away [3].

Exposed to high temperatures, MIC breaks down to hydrogen cyanide (HCN) [19]. At +200°C, three percent of the gas is hydrogen cyanide. At +400°C, the proportion has increased to 20 percent. It has been suggested that alternatives to chronic cyanide poisoning exist, for example, modification of the haemoglobin molecule [19]. This issue became an important controversy. Chloroform may inhibit the breakdown of MIC to hydrogen cyanide [9].

After the Bhopal leakage, UCC as well as some scientists maintained that MIC could not penetrate the lung-blood barrier, and that MIC could not lead to permanent injuries. They said that MIC would neutralise itself immediately in the presence of moisture [20].

MIC is soluble in and reacts with water, e.g., in mucous, and thus can penetrate tissues in the respiratory tract and stomach and interact with proteins. ICMR (Indian Council of Medical Research) found MIC trimers, hydrogen cyanide and other compounds in autopsy samples and maintains that this is evidence that different substances had entered the body's system. A study on methyl carbamylation in post-mortem blood supports this [20].

In a series of experiments, animals were exposed to MIC by inhalation [21]. This caused serious injuries to the entire respiratory tract, the viscera and the brain, as well as genotoxicity. Exposure to methylamine and N,N'-dimethyl-urea, hydrolytic derivatives of MIC, did not have the same effects. Rats exposed to MIC by inhalation or by subcutaneous inception suffered serious damage to the lungs and other organs, although the effects were not exactly the same [22]. The exposure of pregnant rats to MIC resulted in increased numbers of dead foetuses, decreased neonatal survival rates and a higher incidence of foetal malformations [23].

In an article published in 1988, it is described how mice inhaled radioactive MIC. After a few hours, radioactivity was found in most organs, including the foetus. The cause for the effective absorption and distribution is probably that MIC is bound to different proteins [17].

National Institute of Working Life has done a review of the studies on animals [17]. The dose-response and dose-effects curves seem to be steep. The effect on mice and rats has been temporary irritation of airways, necrosis of epithelia, haemorrhagia and oedema of the lungs, and also fibrosis in the bronchi and proliferation of the connective tissue in the lungs. Some effects on the immunosystem were shown, probably secondary to the lung damage.

MIC can inhibit the mitochondria in the breathing chain and thus induce histotoxic hypoxia. MIC also has anticholesterase activity [18] but does not seem to have mutagenic activity. It might be genotoxic by binding with nuclear proteins. Chromosome aberrations in hamsters are shown. The carcinogenic potential is considered low. No conclusions concerning the cause of the effects on foetuses can be drawn – it may be a direct effect of MIC or secondary, due to the mothers' lung injuries.

7.2.4 Other released substances

The information here is mainly collected from the textbooks on toxicology by Patty [24], Clayton & Clayton [25] and Shelley et al [8]. Information on hydrocyanide also comes from Montelius [26].

7.2.4.1 Methylamines

Methylamines have a fish-like odour and are strongly alkaline. Many are capable of cutaneous hypersensitisation. Some have physiological or pharmacological effects such as histamine liberation and vasodilatation. A number of aliphatic amines have been identified as normal constituents of mammalian and human urine. These include methylamine, dimethylamine, and trimethylamine. The amines are easily absorbed by the gut and the respiratory tract. The simple aliphatic amines can produce lethal effects by percutaneous absorption.

Animals exposed to concentrated vapours exhibit signs and symptoms of mucous membrane and respiratory tract irritations. Single exposures to near-lethal concentrations result in tracheitis, bronchitis, pneumonitis and pulmonary oedema. For many of the amines, a single skin application will cause deep necrosis, and a drop in a rabbit's eye results in severe corneal damage or complete eye destruction. Some of the effects observed result from the local corrosive action of the gases in the gastrointestinal tract. Methylamines are able to adversely affect foetal development in mice.

For human beings, local action is primarily irritative and sensitising. Vapours of the volatile amines cause eye irritation with lacrimation, conjunctivitis and corneal oedema, which result in seeing "halos" around lights. Inhalation causes irritation of the mucous membranes of the nose and throat, and lung irritation with respiratory dermatitis. Direct local contact with the liquids is known to produce severe and sometimes permanent eye damage, as well as skin burns. Cutaneous sensitisation has been recorded. Systemic symptoms from inhalation are headaches, nausea, faintness and anxiety. These systemic symptoms are usually transient and are probably related to the pharmacodynamic action of the amines.

Recent research indicates that methylamine could cause mucous membrane and respiratory tract irritation, leading to respiratory depression [22]. A single exposure to a near-lethal concentration results in tracheitis, bronchitis, pneumonitis and pulmonary oedema.

Trimethylamine, one of the degradation products of MIC, has been reported to produce selective growth retardation of male progeny of mice, associated with a decrease in serum testoserone [27].

Methylamine is known to react with nitrates or nitrites, normally present in lake waters, to form nitrosamines, which are known cancer-causing agents [16].

7.2.4.2 Monomethylamine

Monomethylamine (MMA, chemical formula $CH_3 NH_2$) produces the same acute symptoms as MIC, but is much less toxic. MMA is a severe skin and eye irritant and may produce sensitisation. Exposures to the vapours produce eye irritation with lacrimation, conjunctivitis, and corneal oedema. Inhalation causes irritation of the mucous membranes of the nose, throat and lung irritation with respiratory distress and coughing. In animals, single exposures to near lethal concentrations result in tracheitis, bronchitis, pneumonitis, and pulmonary oedema. Its routes of entry include inhalation, ingestion, skin absorption, eye, and skin contact.

MMA has been assigned a health hazard rating of four (the maximum) in the UCC hazard signal system [15].

Due to its transformation into monomethylamine, MIC can itself produce "cherry-red blood" [23].

7.2.4.3 Hydrogen cyanide

Hydrogen cyanide (HCN) and its salts are among the most rapidly acting of all poisons. A few inhalations of higher concentrations (300 ppm) of HCN vapour may be followed by almost instantaneous collapse and cessation of respiration. It inactivates the enzymes necessary for the transport of oxygen into the cells and is capable of suddenly bringing to a halt practically all cellular respiration. The most specific pathological finding in acute cases is the bright red colour of venous blood.

HCN's boiling point is 25.7° C. The maximum safe limit for prolonged exposure is 20 ppm.

Hydrocyanic acid has a characteristic smell of bitter almonds already in concentrations about 1 mg HCN/m³ of air. It can be absorbed by the lungs, by the digestive tract and by the skin.

In cases of severe poisoning with hydrocyanic acid (absorption of a high dose) the victim soon loses consciousness and gets cramps, the pulse and respiration become accelerated and irregular, and death is a question of a few seconds. In less severe cases, first of all breathing difficulties set in, there is breast pressure, tears, a burning ache in the throat and a cough, and photophobia. Later come a feeling of fear, heat, reddening of the face, humming in the ears, and headaches. These symptoms are accompanied by vomiting and diarrhoea. Later on, cramps and unconsciousness set in, followed by respiratory blockage.

If the concentration of the cyanide ion is not so great as to cause death, then with the help of a transulphurase enzyme it is gradually converted into the less toxic thiocyanate and excreted in the urine [24]. The toxicity of thiocyanate is significantly less than that of cyanide. The half time for excretion is 3–9 days. The excretion rate for humans is slower than for rodents or dogs, and is dependent on nutrition and the functioning of the kidneys.

Cyanide can accumulate in the body, if the absorption exceeds the excretion rate, and gives delayed acute or chronic poisoning. It seems like repeated exposure is necessary for this to happen. It has been suggested that chronic cyanide poisoning may be identical to thiocyanate intoxication. Common symptoms are headache, dizziness, nausea and weakness. Less common are rash, increased sweating, dyspnoea, weight loss and irritability. Many other unspecific symptoms are mentioned.

Goitre has been associated with chronic exposure, but it is not known if a single exposure to hydrocyanide affects the thyroid. However, in the 1940s, hypothyroidism and thyroid goitre were discovered to be complications of thiocyanate treatment for hypertension [23]. Maternal hypothyroidism can cause serious sequelae. Paresis and neuropatia can be the result of chronic poisoning after eating bitter cassava.

The introduction of excess thiosulfate ions increases the rate of conversion from cyanide to thiocyanate. This can be done by intravenous injections.

The possibility of a "cyanide pool" in the body has been discussed. This would justify treatment with sodium thiosulfate (NaTs) also a long time after the main exposure. Thiocyanate and cyanide exist in an equilibrium through separate metabolic pathways in each direction. The addition of NaTs would push this equilibrium toward the production of thiocyanate. In addition, it would provide an exogenous sulphur source, which may be critical to a person who is malnourished. Liver damage may have reduced available stores of the detoxifying enzyme rhodanese and thus retarded thiocyanate production. Foodstuffs naturally containing cyanide, and contaminated food, may also contribute to an individual's cyanide pool.

7.2.4.4 Carbon monoxide

Carbon monoxide (CO) takes the place of oxygen in the haemoglobin and thus interrupts the normal oxygen supply to the body tissues.

Carbon monoxide is a colourless, odourless gas that is soluble in water. It is absorbed through the lungs, where it enters the bloodstream in the same manner as oxygen. It has a greater affinity for haemoglobin than oxygen and therefore forms a more stable compound. It thus interrupts the normal oxygen supply to the body tissues. The symptoms are headaches, weakness, dizziness, dimness of vision, nausea, vomiting. Collapse, syncope, convulsions, respiratory failure and death may follow high concentrations of carbon monoxide in the blood. Persons suffering prolonged unconsciousness from exposure to carbon monoxide may have permanent ill-effects such as damage to the heart, blood vessels and various visceral organs, but more frequently to the brain and the nervous system as a result of anoxaemia in the brain tissue.

7.2.4.5 Carbon dioxide

Carbon dioxide (CO_2) is a normal constituent of the atmosphere. In high concentrations, it can contribute to oxygen deficiency. The initial effect of the inhalation of excessive carbondioxide is noticed in concentrations of about 2 percent or 20,000 ppm, when the breathing becomes deeper and the tidal volume is increased. The depth of respiration is markedly increased at 4 percent. At 4.5 to 5 percent, breathing becomes laboured and distressing for some individuals.

7.2.4.6 Phosgene

Phosgene ($COCl_2$) is a colourless gas. It was effectively used as a combat gas during the First World War. It decomposes in water to form carbon dioxide and hydrogen chloride. In higher concentrations, it is a severe irritant to the entire respiratory tract. It causes a rasping, burning sensation in the nose, pharynx and larynx. In the moist atmosphere of the terminal spaces of the lungs, complete hydrolysis occurs with irritant effects upon the alveolar walls and blood capillaries. The result is a gradually increasing oedema, until as much as 30 to 50 percent of the total blood plasma has accumulated in the lungs, causing "dry land drowning". The end result may be either asphyxiation or heart failure, which may be delayed. High concentrations of phosgene are immediately corrosive to lung tissue and result in sudden death by suffocation. Sequelae include pulmonary scarring, lobular emphysema, small irregular areas of atelectasis, and bronchitis.

7.2.4.7 Nitrogen oxides

Nitrogen (N_2) is a colourless, odourless, physiologically inert gas that normally constitutes about 78 percent of the atmosphere by volume. It is soluble in water. The only physiological effects due to inhalation of nitrogen result from oxygen dilution.

Nitrous oxide (N_2O) is a colourless gas, which has a sweetish taste. It is soluble in water. When it is mixed with air, it acts chiefly as an asphyxiant by lowering the oxygen percentage.

Nitric oxide (NO) is a colourless gas slightly heavier than air. It is soluble in water. Nitric oxide is not an irritant, but in animals it has been found to act upon the central nervous system to produce paralysis phenomena and convulsions. It combines with haemoglobin and this is oxidised by oxygen in the blood to methaemoglobin, with resultant anoxia.

Nitrogen dioxide (NO_2) is a dark chocolate-brown gas. It is about four or five times as toxic as NO, and can produce symptoms similar to phosgene. In the alveoli, it reacts slowly with the humid air and the moist surfaces. The result is pulmonary oedema that may lead to symptoms as long as eight hours after the exposure.

Inhalation of nitrous gases can cause bronchiolitis obliterans in the lungs [28].

7.2.4.8 Nitriles

NO_2 reacts with water to form a mixture of nitrous acid (HNO_2) and nitric acid (HNO_3). The nitriles are readily absorbed by all routes. Many of them display toxicological effects that appear to be related to cyanide toxicity. However, not all nitriles dissociate readily to produce cyanide; thus various additional toxicological effects can be noted for specific nitriles. The release of cyanide from some nitriles, such as acetonitrile, may be relatively slow. The toxicity of the individual nitriles differs sufficiently to discourage considering them collectively as "organic cyanides".

7.2.4.9 Chloroform

Chloroform ($CHCl_3$) is a colourless liquid with a boiling point 61.26° C. It was earlier used as an anaesthetic. Chloroform is very rapidly absorbed by the lungs and is widely distributed in the body. High concentrations result in narcosis and anaesthesia because of the depression of the central nervous system. Exposure to lower concentrations leads to inebriation and excitation. Vomiting and gastrointestinal upsets may be observed. Responses to acute exposure have been indicated in the summary as central nervous system depression, liver or kidney injury, and possible cardiac sensitisation.

Chloroform may inhibit the breakdown of MIC to hydrogen cyanide [9].

7.2.4.10 Hydrogen chloride

Hydrogen chloride (HCl) is a gas with great solubility in water. It is extremely irritating to the nose and throat. When inhaled in high concentrations, the gas causes necrosis of the tracheal and bronchial epithelium as well as pulmonary oedema, atelectasis, emphysema, and damage to the pulmonary blood vessels. Lesions of the liver and other organs may occur. Exposure of the skin to gaseous hydrogen chloride, escaping from leaks in apparatus or piping, has caused severe burns.

7.2.5 Hypoxia and Asphyxia

7.2.5.1 Oxygen deficiency in air

When the concentration of oxygen in the atmosphere falls below 16 percent, symptoms of anoxia begin to appear.

Stage	O_2 vol%	Symptoms or phenomena
1	12–16	Breathing and pulse increase, muscular co-ordination disturbed
2	10–14	Disturbed respiration, abnormal fatigue upon exertion
3	6–10	Nausea and vomiting, inability to move freely
4	Below 6	Convulsive movements, gasping, respiration and heart stop

Mixtures of 2 percent oxygen with nitrogen have been administered for three or four minutes in the treatment of certain forms of insanity with only an occasional respiratory failure. There is almost immediate loss of consciousness, progressive stimulation of respiration, tachycardia and irregularity of heart action, muscular twitching, and opistotonos.

The suddenness with which oxygen-deficient atmospheres can cause unconsciousness and death may be explained as follows: When a subject inhales normally and the atmosphere is oxygen depleted, the atmosphere in the lungs is depleted of its oxygen by ventilation much more rapidly than by absorption. Since the initial effects of anoxia are increased rate of breathing and circulation, these processes are speeded up, the oxygen percentage in the lung atmosphere falls below 10 percent, and the arterial blood supply to the brain very quickly (in seconds) becomes insufficiently oxygenated to maintain consciousness.

Whenever anoxia is prolonged, recovery is slow and there are apt to be sequelae such as hallucinations, excitement, headaches, nausea and apathy extending over several hours; these are thought to result from the pressure of cerebral oedema. When the anoxia is severe and prolonged, with unconsciousness, irreversible degenerative changes may occur in the nervous system, especially in the cerebral cortex and basal ganglia. These result in paralyses, amnesia, and other manifestations of permanent injury. The condition is perhaps more often seen following prolonged unconsciousness due to carbon monoxide poisoning.

7.2.5.2 Chemical asphyxia

The oxidation of haemoglobin to methaemoglobin makes it incapable of its usual functions of transporting oxygen, and chemical asphyxia may be said to have occurred. When 60 percent or more of the haemoglobin has been transformed, symptoms of hypoxia may follow. Death due to anoxia can be caused by methaemoglobinemia, but only with very large doses of toxic agents or in highly susceptible subjects.

Various aromatic amino and nitro compounds, nitrites, nitrates, chlorates and quinones can cause methaemoglobin formations. The major manifestation of metha-emoglobinemia is cyanosis, a purplish blue colour of the skin.

Carbon monoxide and hydrogen cyanide also lead to chemical asphyxia (see 7.2.4.3, 7.2.4.4).

7.2.6 Comments

It is probable that the cloud consisted at least of hydrogen cyanide, phosgene, MIC, carbon monoxide, hydrogen chloride, nitrous oxides, monomethyl amine and carbon dioxide.

The sudden deaths without pulmonary oedema can be explained by hydrogen cyanide and/or phosgene in high concentrations, carbon monoxide and oxygen deficiency. The acute deaths with pulmonary oedema could have been caused by MIC, hydrogen chloride and nitrous oxides. Many of the symptoms experienced by the victims at Hamidia hospital were the same as those described for "less difficult cases" of hydrogen cyanide poisoning. The less serious symptoms that were experienced at some distance from the plant could have been caused by MIC and monomethyl amine.

The fact that children were worse affected than adults might be due to the higher concentrations of toxic compounds close to the ground, in combination with higher susceptibility.

The somniferous effect could have been caused by chloroform.

The delayed pulmonary symptoms, after 24–48 hours, might have been caused by phosgene, nitrogen dioxide and nitriles.

Deaths within days/weeks without delayed pulmonary symptoms might have been caused by hydrogen cyanide in low concentrations, MMA in high concentrations, nitrogen oxides, nitriles, middle and low concentrations of phosgene, hydrogen chloride and MIC.

The chronic symptoms from lungs, eyes, intestinal tract and the nervous system as well as the late deaths could have been the result of persistent damage caused by MIC, carbon monoxide, nitrogen dioxide, hydrogen chloride, monomethyl amine, low concentrations of phosgene and hydrogen cyanide, and oxygen deficiency. Liver damage can be explained by hydrogen chloride, anoxia and probably also MIC. Many of those substances are also skin irritants and can cause sensitisation.

It is also possible that the residents reacted more seriously to the gases because of sensitisation to MIC during previous leaks (see 6.3.5).

7.2.7 References

1. Lapierre, D. and J. Moro, *It Was Five Past Midnight in Bhopal*. 1st Indian ed. 2001, New Delhi: Full Circle Publishing. 376.
2. Fera, I., *The Day After*, in *Bhopal: Industrial genocide?* 1985, Arena Press: Hong Kong.
3. Jones, T., *Corporate killing. Bhopals Will Happen*. 1988, London: Free Association Books.
4. *Description on Possible Compounds of the Gas Cloud.*
5. Chandra, H., et al., *GC-MS identification of MIC trimer: a constituent of tank residue in preserved autopsy blood of Bhopal gas victims*. Med Sci Law, 1991. **31**: p. 294–8.

6. Chandra, H., et al., *Isolation of an unknown compound, from both blood of Bhopal aerosol disaster victims and residue of tank E-610 of Union Carbide Limited. Chemical characterization of other structure.* Med Sci Law, 1994. **34**: p. 106–10.

7. *Methyl Isocyanate. Union Carbide F-41443A - 7/76.* 1976, Union Carbide Corporation: New York.

8. Shelley, H., et al., *The health effects of methyl isocyanate, cyanide and monomethylamine exposure,* in *The Bhopal Tragedy,* W. Morehouse and A. Subramaniam, Editors. 1986, The Council on International and Public Affairs: New York.

9. Varadarajan, S.E.A., *Report on Scientific Studies on the Factors related to Bhopal Toxic Gas Leakage.* 1985, Indian Council for Scientific and Industrial Research: New Delhi.

10. Subramaniam, A., *Bhopal - The dangers of diagnostic delay.* Business India, 1985 (August): p. 12–25.

11. Kumar, C.D. and S.K. Mukerjee, *Methyl Isocyanate: Profile of a killer gas,* in *Bhopal: Industrial genocide?* 1985, Arena Press: Hong Kong.

12. Dhara, V.R., *Health Effects of the Bhopal Gas Leak: A Review.* Epidemiol/Prev, 1992. **1992** (14): p. 22–31.

13. Morehouse, W. and A. Subramaniam, *The Bhopal Tragedy. What really happened and what it means for American workers and communities at risk.* 1986, New York: The Council on International and Public Affairs.

14. *How Dangerous are the Icocyanates?,* in *Aktuell Arbetslivsforskning.* 1997, Swedish Council for Work Life Research: Stockholm. p. 3–15 [In Swedish].

15. *Carbon monoxide, Phosgene and Methyl isocyanate. Unit Safety Procedures Manual.* 1978, Union Carbide India Limited, Agricultural Products Division: Bhopal.

16. *The Trade Union Report on Bhopal.* 1985, ICFTU-ICEF: Geneva, Switzerland.

17. *Scientific Basis for Threshold Limit Values. Methylisocyanate and Isocyanacid.* 2001, National Institute for Working Life: Stockholm.

18. Singh, S., *Chemistry, Fate, Pharmacology & Effects of Methyl isocyanate. Only for medical profession.* 1985, Department of Pharmacology, Gandhi Medical College: Bhopal.

19. Kulling, P. and H. Lorin, *The Toxic Gas Disaster in Bhopal December 2–3, 1984.* 1987, Forsvarets Forskningsanstalt (National Defence Research Institute): Stockholm, Sweden. [In Swedish]

20. Sriramachari, S., et al., *GC-NDP and GC-MS analysis of preserved tissue of Bhopal gas disaster: Evidence of methyl carbamylation in post-mortem blood.* Med Sci Law, 1991. **31** (**4**) (Oct): p. 289–93.

21. Sarangi, S., *Methyl isocyanate (MIC),* Bhopal People's Health and Documentation Clinic (Sambhavna): Bhopal.

22. Jeevaratnam, K. and S. Sriramachari, *Acute histopathological changes induced by methyl isocyanate in lungs, liver, kidneys & spleen of rats.* Indian J Med Res, 1994. **99**: p. 231–5.
23. Mehta, P.S., *et. al. Bhopal tragedy's health effects. A review of methyl isocyanate toxicity.* JAMA, 1990. **264**: p. 2781–7.
24. *Industrial Hygiene and Toxicology.* 2nd revised ed, ed. F.A. Patty. Vol. 2. 1963, New York: Interscience Publishers, John Wiley & Sons.
25. *Patty's Industrial Hygiene and Toxicology.* 3rd revised ed, ed. G.D. Clayton and F.E. Clayton. Vol. 2 A. 1982, New York: Wiley-Interscience Publication, John Wiley & Sons.
26. Montelius, J., ed. *Scientific Basis for Threshold Limit Value.* Work and Health. Vol. 19. 2001, National Institute for Working Life: Stockholm, Sweden. [In Swedish].
27. Ranjan, N., et al., *Methyl Isocyanate Exposure and Growth Patterns of Adolescents in Bhopal.* JAMA, 2003. **290**(14): p. 1856–7.
28. Haeggstrom, F., A. Eklund, and G. Elmberger, *Distinguish between BO and BOOP! Two pulmonary diseases that often are mixed up.* Lakartidningen, 1997. **94**(1287–91 [In Swedish].

7.3 Outside the Plant

7.3.1 Time schedule

The residents of Bhopal learned about the gas leak at different points in time:

Table 4 Time schedule for events outside the plant

Time	Ref.	Event
23.30	1, 2	• Gas first noticed by residents in the settlements surrounding the plant.
00.30?	3	• The plant siren was heard for ten minutes. • A pungent odour was noticed.
00.45	4	• People in JP Nagar woke up feeling asphyxiated.
01.00	No ref.	• The plant siren was heard for a few minutes (see 7.1.1).
	5	• The alarm reached the fire corps on the direct telephone line. When they arrived at the plant gates, the plant staff said that "It was just a gas leak" and there was no need for help from the fire corps. The fire corps returned to the station.
	6	• The police superintendent was roused from sleep and told that people were fleeing from their homes.
	7	• A town inspector on patrol sent a wireless message to the police control room.
01.15	4	• The police control room was informed that there had been a major gas leak and that people were fleeing the area. The police contacted UC immediately but were informed by the works manager that even if there had been one, it could not be from UCIL as the entire plant had been shut down.
01.25	7	• The Police Superintendent reached the control room and found the staff coughing

Continues...

...Continued

Time	Ref.	Event
		violently. He put some men on to calling the UC factory.
	7	• Between 01.25 and 02.10 they reached the UC factory three times. They got the answers "Everything is OK" or "We do not know what happens, sir".
01.30	3	• Cattle vomited, their eyes burned and teared, and they died.
01.45	7	• The Additional District Magistrate got through to the UC manager at home. His reply was: "The gas leak just can't be from my plant."
	5	• The Chief Doctor at JP Hospital was informed and left for the hospital. He observed large numbers of people on their way to the hospital. He could smell MIC and his eyes filled with tears, although he was 5 km away from the plant.
02.00	No ref.	• The plant siren was heard (see 7.1.1).
	5, 8	• The army hospital was informed by workers at JK Straw Products. Ambulance and other vehicles were sent in the direction of the plant.
	5	• The first injured people arrived at Hamidia Hospital.
	7	• Up to 02.10, the Police Superintendent got through on telephone to the UC factory three times. Twice he was told "Everything is OK". The third time, "We don't know what has happened, Sir," before the phone was banged down.
02.15	4	• UCIL's public siren sounded. A little later, an engineer from the plant walked into the police control room to announce that the "leak has been plugged".
	5	• The Mayor was informed and left for the plant. He saw a greyish fog that he had to drive around.

Continues...

...Continued

Time	Ref.	Event
02.30	4	• The police decided to evacuate the city.
03.00	7	• The UC management informed the police officials that the leak had been plugged. This was the first official message from the UC.
	4	• The first deaths were reported to the police.
03.45	5	• The Superintendent of Hamidia Hospital arrived there, after having had some difficulties because of the gases.
04.00	5	• The gas concentration had decreased so that a doctor from JP Hospital could enter the affected area.
06.00	8	• Police vans with mounted loudspeakers announced: "Something had gone wrong somewhere. Everything is normal now. Citizens are requested to return to their homes."
06.30	5	• The Chief Doctor from JP Hospital visited the affected area.
Dawn	8	• Private vehicles started to carry people to the hospitals. • The official machinery got into gear and commenced relief operations.
Morning	8	• The Works Manager told press reporters that the leak had been sealed within minutes.
14.00	3, 8	• The first funeral pyre was lit.

7.3.2 The spreading of the gases

The meteorological conditions that night were very unfavourable, aggravating the consequences of the release [9]. There was a soft wind from the north of 1–2 m/s, in the direction of the settlement in the vicinity of the plant. The temperature was 7–10° C and the air was relatively dry. An inversion reduced dilution of the gas cloud. The terrain in the affected area is essentially level, with a small declination towards the railway station and the Old Town.

The cloud, which was actually an aerosol, slowly spread out over the Old Town of Bhopal and later over parts of the New Town and the lakes, according to witnesses (appendix 1). Karlsson et al stress the considerable width and irregularity of a one-hour-average plume as compared to an instantaneous plume, especially if the wind speed is low as in Bhopal [9].

The plume initially travelled westward at a considerable height above the ground. Thus, although the residential areas that lay west of the plant were the first to be hit by the gases, they were spared its highest concentrations [4]. There appears to have been a change in wind direction, which caused the gas cloud to drift southwards, toward the city, where it descended in a heavy mist upon the shantytowns that lay to the south and southeast of the plant.

If such a cloud were released at ground level, the lighter elements would quickly rise, leaving only the heavier ones close to the ground. But the gases escaped from two points: at ground level, from the caustic overflow tank, and at 100 feet (30 m), from the atmospheric vent at the top of the MIC tower structure. The heavier gases descending from the atmospheric vent trapped the lighter elements released at ground level, and held them close to the ground [4].

The area with the largest number of casualties was 6–7 sq.km. south of the factory [9]. The region with severe, but not deadly, injuries was about 25 sq.km. The railway station is situated roughly in the middle of the affected area.

People within a sector of 150–180 degrees suffered severe poisoning injuries. Lethal injuries were found up to a distance of 2.5 km and severe, but not lethal, injuries occurred up to about a distance of 4 km. Others estimate that the gas spread far enough to seriously injure people within eight kilometres downwind [10].

7.3.3 In the bastis

The situation outside the factory is best described by Lapierre and Moro [3].

Sunday, the 2nd of December, was the first day of Isthema, the great Muslim prayer conclave. Thousands of pilgrims had arrived in Bhopal. It was also the wedding season, with relatives coming from all over India. At the railway station, all the staff, including 101 *coolies*, was gathered, waiting for the train that was delayed by the thick fog.

The residents of Bhopal who were not celebrating were generally asleep, on the streets or in the station, in *kuccha* houses without doors or windows, in *pucca* (permanent) houses with windows and doors, or on the second or third floor in the Old Town. Many of those who lived nearest to the plant died in their sleep. Most woke up because they were coughing and suffocating. Then they felt as if they had swallowed something like "burned chilli"; their eyes started to burn as well as their respiratory passages, and they began to vomit. When they looked outside, they saw a white mist. Some stayed in bed under a blanket, but most people went out, scared and angry, and tried to get away from the cloud. Some died instantaneously. The others ran, or used vehicles if possible, and moved away from the factory, following the direction of the cloud. Being blinded, they shouted for their family members – but soon their throats were constricted by the gas, their lungs choked. As they ran, they inhaled larger amounts of the gases. Later, they tried to get to the hospitals. Many never reached them.

Those who were left or stayed behind were often better off, as they inhaled smaller amounts of the gases. A few went the opposite way, against the wind, and managed better.

The two railway employees who were on shift that night stuck to their posts and tried to send messages to adjoining stations to stop all incoming trains. When the Gorakhpur Express approached, they walked onto the rails and signalled with lamps – but were not detected until it was too late. When the train arrived at the station, the stationmaster immediately went out to send it away. The two employees died, and the station master became an invalid.

When the morning light came, the extent of the disaster was obvious. In the areas around the factory, every goat, cat, dog, cow and buffalo had died. Outside and inside the houses, lay human bodies. Only the birds and rats did not die. Within a few days, all the leaves of the trees fell off, and the grass turned yellow.

The station was, as usual, crowded with travellers, porters, homeless people and also a group of gypsies. They were all found dead next morning.

On December 4, a number of police trucks entered the area next to the factory. The dead bodies were loaded onto the trucks and dumped into the river. One or another was not dead, but woke up in the cold water, surrounded by dead bodies. We will never know how many people were burned or buried alive.

During the first days after the disaster, the citizens of Bhopal rushed to help [8]. Hundreds of volunteers, including students, used their own vehicles and took victims to the hospitals, removed dead bodies, assisted at funerals, and brought cooked food to hospitals and medical camps.

The authorities, confronted by a health crisis of catastrophic proportions, had to clear the city of the dead. This included 2,000–4,000 heads of cattle and several hundred dogs, cats and birds. The voluntary agencies played an important role in this. No one was bothering to count the bodies at this stage.

The day after the leak, several thousand Bhopal residents tried to storm the factory. Plant officials and police guarding the plant only succeeded in turning the crowd away by telling them that another poisonous gas leak was in progress [11].

7.3.4 At the hospitals

The doctors and the staff were completely taken by surprise as thousands upon thousands of survivors entered the gates [3]. They were half-blinded, gasping for breath, foaming at the mouth and vomiting. Most of the patients had to be treated outside. There was no time to keep any kind of records. All types of medicines were tried to give relief. When the hospitals got hold of the doctor of the plant, they were given the message: "It is only like tear gas".

The patients' clothes, hair and beards were impregnated with toxic emissions that also affected the medics. One student died after successfully treating a child with mouth-to-mouth respiration.

Many dead bodies were taken to hospital for registration and identification. At the JP Hospital, the authorities were told not to register more than 550 dead persons (personal communication). Tapes were attached to the forehead, the number and day noted. When a figure close to 550 was reached, the bodies were put on trucks and taken for a mass burial.

The Bhopal hospitals treated 130,000 patients, and 40,000 were treated in other hospitals in Madhya Pradesh [12]. Of those, 12,000 were in a very critical condition and patients at the hospitals had respiratory difficulties, with pulmonary oedema, and eye problems. Vomiting was also a common complaint.

As many as 700 doctors, including 200 from outside Bhopal, were attending to the gas-affected patients by the evening of December 3 [8]. They were assisted by 250 nurses, 870 para-medicos and volunteers. Students, including those from the local medical college, came forward to render assistance in a big way. Medical experts from all parts of the country and abroad rushed to Bhopal to guide and assist in the treatment of the patients. Seven hospitals, 26 dispensaries, and 14 additional temporary dispensaries worked round-the-clock.

7.3.5 Other actors

As the police station was situated in the affected area, the work could not be organised from there [5]. The co-operation of the police was sporadic and has come in for criticism. However, they blockaded roads and railways to incoming traffic.

Around a thousand military personnel took part in the resuscitation work, mainly by transporting injured persons. One soldier died and 20 were injured by the gases.

Public buses as well as private cars and taxis transported people to other towns. It is estimated that 20,000 people left the town [5]. Some serious traffic accidents occurred.

Volunteers at Hamidia hospital hunted for medicines and other equipment at chemists' and other shops.

7.3.6 The environment

The ambient temperature in Bhopal went up from 13 to 34 degrees Celsius the day after the explosion [10].

The gases immediately caused visible damage to the trees surrounding the factory, changing their colour and blackening the leaves [12]. Within a few days, all the leaves fell off. Leafy vegetables such as cabbages showed white spots on their green leaves. In short, the vegetable crops all showed signs of being badly affected. Damage to trees was visible 1.5 km from the factory [9].

For several days, or even weeks, gas was trapped in cupboards and closed rooms. The ground, the houses, clothes and the water were covered with a layer of white powder that later turned green.

Tests conducted on the initiative of science students indicated the presence of MIC as high as 0.15 ppm as late as December 5 [13]. Air samples taken on December 5-6, 1984, by the Pollution Control Board, showed 4.5 ppm cyanide in the air near the MIC storage tank, indicating a continuing leak of cyanide from the tank [11].

Scientists were calling for affected areas to be evacuated, since the gas was likely to remain in the atmosphere for three or four weeks. It was suggested that the atmosphere in the affected areas should be sprayed with diluted ammonia [11].

A team from the Council of Scientific and Industrial Research (CSIR) undertook an examination of the environmental impact in the days following December 5 [14]. The tests carried out did not show any presence of MIC or related toxic materials in the environment (though the test results were never released). They recommended that the Government advise the public to wash all vegetables and food articles with water and to clean floors, walls and surfaces with water.

On December 3, raw water from the lakes and filtered water were tested by the Public Health Engineering Department, and the drinking water was declared safe. The test results were not presented [11].

Studies of bodies of water in the severely affected areas show fish mortality rates up to 70% [6, 15], but analyses were never made [16]. On December 8, the main fish market was sealed by the authorities, and the residents were recommended not to eat fish [8].

A related problem was taking care of the nearly 2,000 bloated animal carcasses. Dumpers and cranes were used, and they were carried 5 km away where bulldozers had prepared an enormous grave. Salt, bleaching powder, lime and caustic soda were first spread on the bottom, because of the risk that the bodies would burst [8].

7.3.7 The second exodus

7.3.7.1 Operation faith

"Operation Faith" is described in the CSIR report [14], by Jones [11] and by Lapierre and Moro [3]. It was estimated that about 15 tonnes of MIC was present in tank 611, and there were some residues in tank 619. The chairman of UCC in USA sent telex messages asking the Government of India to proceed with the processing immediately, as the risk of gas leakage was increasing every day.

The former American chief engineer was in charge of the operation. Thorough preparations were made, before the residues were converted to Sevin. The safety precautions inside the factory area were rigorous. The safety systems were repaired. A communication system and risk management control organisation were developed. This included bringing back UCIL operators into function and giving them training to work under unusual conditions.

The date for Operation Faith was decided to be December 16, a Sunday. The plans were made public three days in advance. Radio and television were used for disseminating information on precautions to be observed.

The plan was to evacuate about 80,000 persons living in relatively open structures to safe places. All educational institutions were closed. An alarm system was established by the Government to alert the public of any imminent leakage. The public was advised to go into closed buildings when an alert was sounded and to use wet towels on their faces to filter air for breathing.

But the people of Bhopal did not trust the authorities. They did not find the sites chosen by the government safe. They left the town on all kinds of vehicles, with all their belongings. The elite were the first to leave.

The amount of MIC processed during Operation Faith was 22 tonnes, nearly 50 percent more than the UCC experts had estimated as remaining in the plant.

7.3.7.2 The third exodus

In March 1985, the government began to move chemicals out of the factory. Due to a chlorine leak, three workers were injured [10]. When chlorosulphuric acid fumes leaked, hundreds of slum-dwellers fled the area.

7.3.8 Information

7.3.8.1 Information from UCC

In the material studied, it has been very difficult to find facts about what information Union Carbide Corporation and Union Carbide India Limited provided to workers, authorities, health care services and residents. Instead, complaints of a lack of information or misinformation are legion.

Not even the medical doctor of the Bhopal plant had proper information about the properties of the gases. He continued to insist that MIC was only an irritant and not life-threatening [12]. In reply to telegrams sent to UCC's US headquarters, doctors in Bhopal were told that the gas was "harmless" [10].

The first information on the nature of MIC reached the medical authorities from a newspaper report. Activists contacted Greenpeace, who did a computer search and sent the material they found with Lufthansa to India. In a telex of December 5, UCC briefly outlined possible treatment, but did not describe the possible toxic effects of MIC. Several days after the leak, information from the WHO arrived.

UCC finally provided information to the Government of India "presumably between 3 March and 15 March", three months after the killings. Even this was unsatisfactory: "In

October 1985, Dr Dass* was still complaining that he had not received adequate information from UCC" [11].

An Indian government spokesperson was of the opinion that "Carbide is more interested in getting information from us than in helping our relief work" [11]. Although doctors from the UCC headquarters in Danbury turned up in Bhopal, they did not give any information.

Up till today, UCC has not released information about the possible composition of the cloud.

7.3.8.2 Information to residents

The issues here are also lack of information or misinformation.

On the fateful morning, the All India Radio (AIR) made no relevant announcements, which could assist the panic-stricken people. Even the broadcast schedule of the AIR and Doordarshan (TV-channel controlled by the government) did not change till later in the day. When they did come on the air at the usual time, the tragedy was mentioned only in passing in the regular news bulletin [15].

Eight hours after the leak, the atmosphere in Bhopal was declared free of the gas, though people were warned they should be careful about what they ate for another seventy-two hours [11].

Chauhan maintains that the public was advised to wash all vegetables and food articles with water and clean floors, walls and surfaces with water [8]. The residents, however, do not remember having heard this advice.

Other authors describe that within a few days of the accident, presumably on the basis of preliminary investigations, formal statements were issued that air, water, vegetation and foodstuffs were safe everywhere in the city. At the same time, television features informed people that poultry was unaffected, but warned people not to consume fish, etc. Confusion was rampant and people were asking whether it was safe to consume eggs, vegetables, etc. [15].

It was an open secret that consignments of fresh vegetables and fruit were arriving regularly from the government farm at

---

*   Government of India's commissioner in charge of relief of gas victims

Pachmari, a hill resort near Bhopal, for ministers and senior bureaucrats [11].

7.3.9 Comments

Because of the different properties of the compounds of the cloud, it is likely that people living at various distances from the plant were exposed to different compounds in different concentrations.

This may well explain the concentration of deaths in the area close to the factory, the statements of the witnesses, and the findings made by IMCB, that those who lived closer to the plant had more symptoms and more pathological test results than those who lived further away.

The gases were heavier than air. It is thus likely that children were exposed to higher concentrations of the toxic gases as well as to lower concentrations of oxygen. This would have resulted in higher concentrations per kilogram weight than for the adults, and thus led to the higher death-rate among children [8].

7.3.10 References

1. Kalelkar, A.S. and A.D. Little. *Investigation of large-magnitude incidents: Bhopal as a case study*, in *The Institution of Chemical Engineers Conference on Preventing Major Chemical Accidents*. 1988. London.
2. Chouhan, T.R., *Bhopal: The Inside Story. Carbide workers speak out on the world's worst industrial disaster*. 1994, New York: The Apex Press.
3. Lapierre, D. and J. Moro, *It Was Five Past Midnight in Bhopal*. 1st Indian ed. 2001, New Delhi: Full Circle Publishing. 376.
4. Morehouse, W. and A. Subramaniam, *The Bhopal Tragedy. What really happened and what it means for American workers and communities at risk*. 1986, New York: The Council on International and Public Affairs.
5. Kulling, P. and H. Lorin, *The Toxic Gas Disaster in Bhopal December 2-3, 1984*. 1987, Forsvarets Forskningsanstalt (National Defence Research Institute): Stockholm, Sweden. [In Swedish].
6. Malmberg, C., *The whole third world is poisoned* South Asia - Political and cultural magazine, 1985(3): p. 9-11. [In Swedish].

7. *The Bhopal Gas Tragedy 1984– ? A report from the Sambhavna Trust, Bhopal, India.* 1998, Bhopal People's Health and Documentation Clinic: Bhopal.
8. Chauhan, P. S., *Bhopal Tragedy. Socio-legal Implications.* 1996, Rawat Publications: Jaipur.
9. Karlsson, E., et al., *The Bhopal Catastrophe: Consequences of a Liquified Gas Discharge.* 1985, Forsvarets Forskningsanstalt (National Defence Research Institute): Umea, Sweden.
10. Cassels, J., *The Uncertain Promise of Law: Lessons from Bhopal.* 1993, Toronto: University of Toronto Press Inc.
11. Jones, T., *Corporate killing. Bhopals Will Happen.* 1988, London: Free Association Books.
12. *The Trade Union Report on Bhopal.* 1985, ICFTU-ICEF: Geneva, Switzerland.
13. Ramaseshan, R., *Profit against Safety,* in *Bhopal: Industrial genocide?* 1985, Arena Press: Hong Kong.
14. Varadarajan, S.E.A., *Report on Scientific Studies on the Factors Related to Bhopal Toxic Gas Leakage.* 1985, Indian Council for Scientific and Industrial Research: New Delhi.
15. *Bhopal Gas Tragedy.* 1985, Delhi Science Forum: New Delhi.
16. Bidwai, P., *The poisoned city - diary from Bhopal,* in *Bhopal: Industrial genocide?* 1985, Arena Press: Hong Kong.

7.4 Acute Health Effects and Treatments

7.4.1 How many were affected?

A total of 36 wards were marked as being "gas affected", with an estimated population in 1984 of 520,000. Of these 520,000 exposed people, 200,000 were below 15 years of age and 3,000 were pregnant women.

On the basis of the mortality figures, which became available immediately after the gas leakage, these areas were classified as being severely (two wards with 32,000 inhabitants), moderately (five wards with 72,000 inhabitants) and mildly affected (29 wards with 416,000 inhabitants).

According to ICMR, 521,262 people were affected by the gases [1]. This figure does not include victims who were not permanent residents of Bhopal: pilgrims, wedding guests, people with no permanent address and members of nomadic communities. Nor does it include those victims indirectly affected by the tragedy, such as children still in their mother's wombs, those subsequently born to parents poisoned by the gas, or family members of pilgrims and wedding guests.

However, there are many doubts concerning the accuracy of these figures. The area covered by the cloud was probably much larger than first estimated. In 1989, when decisions about interim relief were taken, all 56 wards in Bhopal were considered to be affected: two severely, five highly and 47 generally affected. Today, more than 14,000 death cases and 730,000 cases of personal injuries have been awarded.

7.4.2 How many died?

The first official death toll figure was 1,408. At 1,754, the reckoning was stopped, considering the amount of compensation that would have to be handed out [1]. Later it rose to 2,259, including 541 children and 318 women [2]. In 1991, 3,928 deaths had been certified. Kulling and Lorin make this estimate [3]: 500 persons died before they got any medical treatment. Of the 6,000 who got treatment for serious symptoms, 2,000 died

Dead buffaloes, 1984 (Sambhavna Trust)

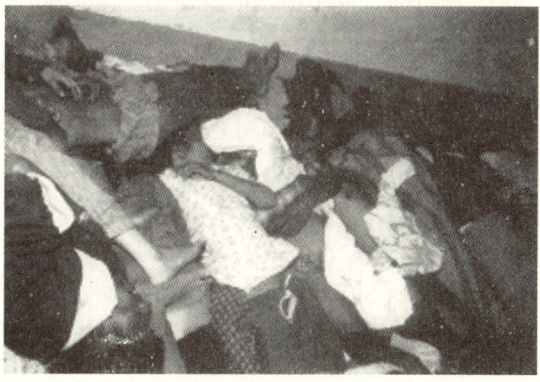

Hamidia Hospital with dead bodies lying around, 1984 (Sambhavna Trust)

within the first week. According to Pandey [4], those who died within hours of the tragedy numbered 2,500, and within months 3,100 more persons died. Morehouse et al [5] considered that there was substantial evidence to support a figure of 5,000. Independent organisations recorded 8,000 dead the first days.
Even this is probably an underestimate.

- Registration at the hospitals was cut off when an upper limit or top number was reached.
- On December 3, only 574 deaths had been registered at hospitals, but the number of corpses received by crematoriums and burial grounds exceeded 1,200.
- As soon as a patient was declared dead, it was common for his/her relatives to vanish with the body, before registration could be done. At Hamidia Hospital, it was estimated that‘ 500 to 1,000 bodies were taken away without registration [5].
- Bodies collected on the streets by the police were dumped in the Narmada River without being registered.
- It is estimated that not even half the dead were buried or cremated in official grounds [5].
- 10,000 shrouds for Hindu and Muslim death services were distributed.
- More than 7,000 corpses were cremated at the Hindu funeral site [1].
- Mass graves have been discovered. In one case, seven bodies were buried, but only one death certificate was issued [5].
- Immigrant workers, gypsies, wedding guests, pilgrims and others with no fixed address in Bhopal, were not accounted for.
- Those who fled Bhopal and died in other places were not registered.
- When a whole family was wiped out, there was no one left to apply for damages.
- In March 1985, there were still thousands of people registered as missing [6].

Other estimations vary between 10,000 and 20,000 [5]. Sambhavna estimates 8,000 deaths during the first weeks, and another 8,000 since then [7]. During the first 48 hours, the death rate in some of the worst affected areas has been estimated at 20/1,000 [8]. During December 1984, it was 24/1,000, compared to the national average of 1/1,000. The worst affected group was children below 5 years age, with a death rate of 33/1,000. JP Nagar, one of the worst affected areas right across the road from the plant, lost at least 25 percent of its 7,000 inhabitants [5].

Around 800 buffaloes and 3,000–4,000 other larger domestic animals died or had to be put to death [3].

7.4.3 Short-term impact on health

The patients who invaded Hamidia hospital had spasms and convulsions. They gasped for breath, their nostrils were quivering, the lips were cyanotic. They had foetid breath and blood-streaked froth. When the doctors listened with their stethoscopes, they heard gurgling rattles from the lungs and faint heart sounds. Other symptoms were spasms in the oesophagus and intestines, attacks of blindness and sweating. The patients were bewildered and had amnesia. Sometimes they behaved like they were mad [1].

The acute symptoms can be described as burning in the respiratory tract and eyes, breathlessness, stomach pains and vomiting. Those living close to the factory had very acute as well as long-term symptoms. Several kilometres away, in the New Town, the residents only felt a passing mild irritation in the respiratory passages and eyes. Those who worked with patients or dead bodies suffered from delayed symptoms, even if they had not been exposed to the leakage as such.

The acute clinical picture was transient irritation and redness of the skin, intense irritation of the eyes including blepharospasm, profuse eyelid oedema and superficial corneal ulcerations [3]. A soothing and somniferous effect was also reported [4]. The respiratory tract physical findings were rhinitis, pharyngitis, coughing, respiratory distress including broncho-constriction, shortness of breath and choking. Many patients died from choking or reflexogenic circulatory collapse. Pulmonary oedema developed in many patients in the acute stage. In others, pulmonary oedema developed later, after a free interval. All types of complications from the respiratory tract were seen, such as pneumothorax, subcutaneous and mediastinal emphysema, bronchopleural fistulas, secondary infections, etc.

Campaign postcard

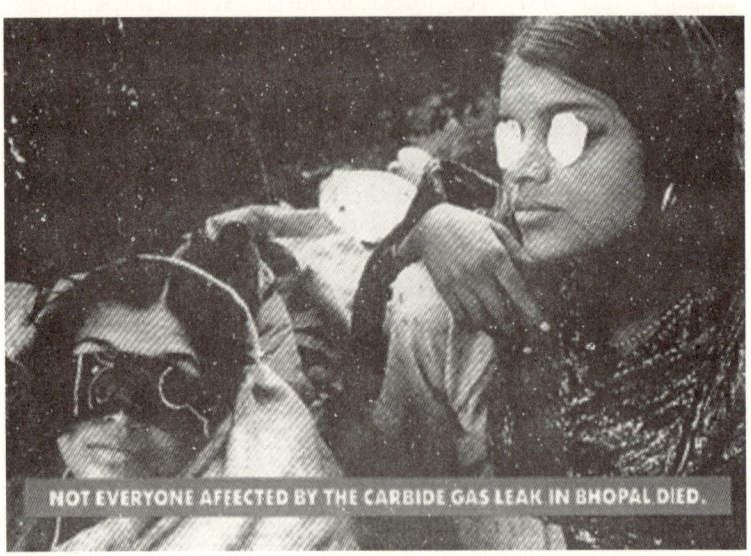

Campaign postcard

The worst hit were children below two years, old people and persons with previous pulmonary diseases, like chronic bronchitis and emphysema [3].

Witnesses say that the dead had greenish coloured skins [1]. They also gave off no smell of decomposition. At autopsies, there was a smell of bitter almonds.

Within barely two weeks of the exposure, Bhopal faced a jaundice epidemic, which doctors suspected was chemically, rather than virally, induced. Other delayed effects reported include intestinal bleeding, pain in the kidneys, general debility. Most of the deaths recorded after the third day involved a failure of the central nervous system [5].

The findings during autopsies on victims revealed changes in many organs, but the most pronounced findings were related to the lungs. The lungs were enlarged and oedematous, showed congestion, haemorrhage and consolidation, with microscopic findings such as bronchiolitis and pulmonary oedema. There were focal haemorrhages in the other organs. In addition, the consistency of the brain was softened through cerebral oedema. The kidneys showed congestion and tubular necrosis. In a large number, the liver showed fatty degeneration. In the gastro-intestinal tract, necrotising enteritis was found.

At Hamidia hospital, nine cases of partial paralysis were found.

The reproductive health of women was affected. Immediately after the gas leak, the stillbirth rate increased by up to 300 percent and the perinatal and neonatal mortality rate by 200 percent. The spontaneous abortion rate increased three to four times and stayed raised for several years. The rate of congenital malformations increased.

A study prepared by ICMR found that the spontaneous abortion rate following the gas leak was 24.2 per cent, about three times the national average [9]. The stillbirth rate was 26.1 per 1,000 deliveries, compared with a national figure of 7.9 per 1,000. A year after the disaster, the infant mortality rate in Bhopal was 110 per 1,000 births, compared with a national average of 65.2 per thousand [10].

UCC claimed that as MIC rapidly hydrolyses in the presence of water, the gas would be neutralised by moisture in the lungs and therefore could not possibly be absorbed.

7.4.4 Treatment

No official line of treatment was forthcoming because of the lack of information. From the morning of December 3, several persons tried in vain to get information from UCIL about the composition of the cloud. It was not possible to reach the WHO or American institutions as it was still Sunday there. The first information whatsoever came by a telex from Greenpeace in Germany.

On December 5, a telegram from UCC stated that "if cyanide is suspected, use amyl nitrite. If no effect, use sodium nitrite 0.3 grams and sodium thiosulphate 12.5 grams." The company later claimed that this advice had been given inadvertently.

A telex message on December 11, 1984, from the Disease Control Centre at Atlanta gave the following information [11]:
• No antidote is known for MIC exposure.
• Give cortisone and oxygen.
• If cyanide poisoning is suspected, it should be treated suitably.

In the acute phase, eyes were treated through irrigation, homatropin, and local antibiotics and in some cases corticosteroids [3]. Symptoms from airways and lungs were treated with oxygen, bronchodilators, diuretics (furosemid) and corticosteroids. While oxygen inhalation at worst proved ineffective, the ingestion of furosemid only made matters worse [5]. When the steroid treatment was stopped after 2–3 days, pulmonary oedema returned in about 40 percent of the patients, and was resolved when steroid therapy was reinstituted. Stomach problems were treated with antacids. Antibiotics and corticosteroids were not available in sufficient quantities [12].

7.4.5 The Detoxification Question

The detoxification issue is described by Kulling and Lorin [3], Sathyamala [13], Jones [14], Morehouse [5] and in the Trade Union Report [6].

Already on December 3, the superintendent of the Medical-Legal Institute in Bhopal found cherry red blood and suspected cyanide poisoning during autopsies. The victims had died of respiratory arrest even in cases where neither pulmonary oedema or the perforation of lung tissues was evident. Nor was there any evidence of cyanosis. Corpses showed little sign of decomposition even after three days of unrefrigerated storage. In some of the examined bodies, traces of cyanide were found, in others methylamine or MIC.

On the same day, the superintendent suggested distribution of sodium thiosulphate (NaTs), a non-toxic antidote to cyanide, to all seriously ill patients. The medical superintendent at Hamidia Hospital and the vice-dean for Gandhi Medical College in Bhopal, however, demanded that cyanide poisoning should be verified chemically in every single case.

On December 5 and 11, telegrams from USA arrived which recommended antidote treatment "if cyanide poisoning was suspected" (see 7.4.4).

A toxicologist from West Germany arrived some days after the accident, bringing 50,000 doses of NaTs. He verified the autopsy findings, and found that blood samples of dead victims contained 2 ppm* cyanide. He conducted a small trial on 50 patients with good effect, but did not get permission to start treatment on a larger scale. He then treated two unconscious patients, of whom one recovered dramatically while the other died. This started rumours that he had killed a patient with NaTs and he had to leave.

On December 13, the chief for Health Services in Madhya Pradesh sent a circular to all doctors in Bhopal not to use NaTs for treatment if cyanide poisoning was not verified through laboratory tests. He stated that he had received this information from ICMR. But ICMR denied this, and in their own circular, there was instruction on how to administer NaTs.

In December 1984, a small double blind clinical trial on the intravenous administration of NaTs was carried out by ICMR. A highly increased excretion of urinary thiocyanate and marked symptomatic relief followed.

Based on the experiences from the trial, NaTs was recommended as the drug of choice, and guidelines were drawn up. Three days after the press statement by ICMR, there was another press statement from the Gandhi Medical College, which

---

* Parts per million

argued that there was no evidence in the medical literature about any form of chronic cyanide poisoning. On the basis of this statement, the MP Government decided not to support NaTs therapy.

Not until January 7, was the first circular of ICMR with instructions on how to administer NaTs available to the doctors in Bhopal.

In April, ICMR released some more information, and the MP Government officially sanctioned the use of NaTs. During the period June 1985 to the middle of 1987, NaTs treatment was given at JSK (People's Health Centre), an NGO clinic.

As an explanation for the increase in the urinary concentration of thiocyanate in survivors, three possibilities were suggested:

- It was a rebound effect after inhibition of the thiocyanate metabolism caused by exposure to MIC.
- It was caused by some other contaminant or diet factor.
- The population was exposed to some cyanide that was released with the MIC.

A number of different alternatives were suggested to account for the cause of chronic cyanide poisoning, for example, modification of the haemoglobin molecule [3, 14]. This issue became a subject of an important controversy.

7.4.6 Comments

7.4.6.1 Mechanisms

It is obvious that most of the organs of the body were affected by the gases and/or hypoxia. Four different mechanisms may be responsible for the damage to the body:
- Direct damage to the mucous membranes of the respiratory tract, the stomach and the eyes;
- Damage to the brain caused by lack of oxygen;
- Toxic effects on cells after penetration of the blood tream, including chromosomal aberrations;
- Post-traumatic stress disorder (PTSD).

7.4.6.2 *Disaster management*

Exposure to the gases would have been reduced if the inhabitants had

- been warned by the alarm earlier;
- covered their faces with a wet cloth;
- stayed inside if they lived in good houses;
- walked instead of ran;
- moved at right angles to the wind instead of moving in the same direction as the wind.

It is likely that exposure to different toxic compounds would have been reduced if the inhabitants had been informed of the necessity to wash their clothes and homes and not to eat or drink contaminated food or water, and if clean food and water had been distributed.

7.4.6.3 *Medical treatment*

Subsequent research has shown that the treatment that is indicated for toxic pulmonary oedema is symptomatic treatment with oxygen, beta-2-stimulators, and when needed, continuous positive airway pressure (CPAP) or positive end-expiratory pressure (PEEP). There are different views on the relevance of treatment with corticosteroids [15, 16]. It is obvious that the doctors in Bhopal gave the best treatment, using the resources they had. Because of having no training in disaster medicine, they might, however, have used the respirators on patients who were already doomed to death.

There are strong indicators that cyanide was present in the gases. It is obvious that sodium thiosulphate, which is a very non-toxic substance, should have been tried as an antidote on a large scale, as soon as possible. There was no real reason not to try it at a later stage. It is extremely possible that impairment would have been mitigated for a large group of survivors if they had been treated with sodium thiosulphate as soon as it was available in Bhopal.

The question of detoxification is one example of the confusion within the scientific and medical establishments that characterised the first year after the leak.

To admit that sodium thiosulphate could give symptomatic relief would be to admit that toxic gases had crossed the blood-lung barrier. This is probably the explanation of why UCC withdrew their recommendation on treatment.

7.4.6 References

1. Lapierre, D. and J. Moro, *It Was Five Past Midnight in Bhopal*. 1st Indian ed. 2001, New Delhi: Full Circle Publishing. 376.
2. Karlsson, E., et al., *The Bhopal Catastrophe: Consequences of a Liquified Gas Discharge*. 1985, Forsvarets Forskningsanstalt (National Defence Research Institute): Umea, Sweden.
3. Kulling, P. and H. Lorin, *The Toxic Gas Disaster in Bhopal December 2–3, 1984*. 1987, Forsvarets Forskningsanstalt (National Defence Research Institute): Stockholm, Sweden. [In Swedish]
4. *Commitments for Sustainable Development. Trade Unions at the Commission for Sustainable Development. Special "Business and Industry" Segment*. 1998, ICFTU: Brussels.
5. Morehouse, W. and A. Subramaniam, *The Bhopal Tragedy. What really happened and what it means for American workers and communities at risk*. 1986, New York: The Council on International and Public Affairs.
6. *The Trade Union Report on Bhopal*. 1985, ICFTU–ICEF: Geneva, Switzerland.
7. *The Bhopal Gas Tragedy 1984– ? A report from the Sambhavna Trust, Bhopal, India*. 1998, Bhopal People's Health and Documentation Clinic: Bhopal.
8. Chauhan, P. S., *Bhopal Tragedy. Socio-legal Implications*. 1996, Rawat Publications: Jaipur.
9. Bhandari, N.R., et al., *Pregnancy outcome in women exposed to toxic gases at Bhopal*. Indian J Med Res, 1990. Febr: p. 28–33.
10. Cassels, J., *The Uncertain Promise of Law: Lessons from Bhopal*. 1993, Toronto: University of Toronto Press Inc.
11. Chandra, H., et al., *GC-MS identification of MIC trimer: a constituent of tank residue in preserved autopsy blood of Bhopal gas victims*. Med Sci Law, 1991. **31**: p. 294–8.
12. *Bhopal Gas Tragedy*. 1985, Delhi Science Forum: New Delhi.
13. Sathyamala, C., *The medical profession and the Bhopal tragedy.* Lokayan Bulletin, 1988(6:1/2).
14. Jones, T., *Corporate killing. Bhopals Will Happen*. 1988, London: Free Association Books.
15. Karlson-Stiber, C., et al., *Toxic pneumonitis in ice-hockey players*. Lakartidningen, 1996. **93**: p. 3808–12 [In Swedish].
16. Lorin, H.G. and P.E.J. Kulling, *The Bhopal Tragedy - what has Swedish disaster medicine learned from it?* J Emerg Med, 1986. **4**: p. 311–6.

THE POST-EVENT PHASE: LONG TERM EFFECTS

8.1 Long-term Impact on Health

8.1.1 General aspects

For several years, there were different opinions about the long-term effects of the leak. UCC consistently maintained that MIC could not cause permanent impairment. Sections of the scientific and the official establishments were of the same opinion. The survivors' organisations, activists and NGOs, however, have fought to have their opinion acknowledged, namely, that thousands of people have been permanently disabled by the gases.

In 1986, Dr C R Krishna Murti, who was the president of the commission that investigated the accident, visited Stockholm [1]. He stated that 30,000–40,000 persons had persistent disabilities. He recognised the following categories:

- Those who were so seriously disabled that they cannot work. They often experienced difficulties in walking or cycling because of bad co-ordination.
- Those who had some persistent dysfunction in the airways and suffered from chronic pulmonary insufficiency, but still manage to work.
- Those who seem well, but who have a strongly decreased resistance to infections, especially in the lungs and airways.

The experiences of witnesses at different distances from the plant differed. Those living close to the factory suffered from very acute as well as long-term symptoms. For some, the symptoms became aggravated with time. Those who lived a little further away, or who spent the night on the second or third storey, recovered better, with only a slight degree of breathlessness remaining. Those who lived on the slopes of the southern part of the Old Town have no remaining symptoms, but they live with the memories and are worried about the effects on future generations.

In their annual reports from 1990 and 1992 [2, 3], the ICMR reports on the long-term studies, in which many long-term health effects were found. The long term symptoms 10–15 years after exposure have been investigated by the International Medical Commission on Bhopal (IMCB) [4–8], Eckerman [9] and Sambhavna [10]. Earlier findings were verified.

In the worst affected *bastis*, Jaiprakash (JP) Nagar, at least one person in every family is acutely ill, and at least two people in every family are under constant medication. There is a tendency for accumulated deaths to occur in certain families.

The survivors complain of breathlessness, coughing, chest pains, fatigue, body aches, abdominal pain, numbness and tingling in the limbs, weak sight and runny eyes, anxiety attacks, bad memory, concentration difficulties, irritability, headache and mental illnesses.

An unusually large number of women have menstrual irregularities and excessive vaginal secretions. Mothers complain of retarded physical and mental growth in children exposed at infancy or born after the disaster.

Symptoms of fever, burning sensations in the body, loss of appetite, numbness and tingling in the limbs, backache, giddiness and *ghabravat* (panic attacks) seem to have manifested 3–4 years after the disaster and are getting worse.

A reasonable estimate is that between 100,000 and 200,000 people are permanently impaired.

The ICMR centre in Bhopal closed down in December 1994, after ten years of work. The Centre for Rehabilitation Studies (CRS) at the MP government has now taken over responsibility for the long-term research. The cohort studies show an overall over-morbidity among the gas affected compared to the control group [11].

8.1.2 Investigated and reported long-term effects

8.1.2.1 General aspects

Many of the epidemiological and clinical studies have methodological shortcomings, which are discussed in section 8.2.

In one study, a cohort of 113 patients was followed up to two years after the gas leak [12]. Symptoms like muscle weakness, loss of memory, depression and anxiety seemed to increase with time, while irritation and concentration difficulties decreased. Symptoms from the respiratory tract first decreased, then again increased. X-ray of lungs showed interstitial changes in lungs and pleurae.

In the IMCB investigation 1994 [13], the exposure of individuals was estimated by developing exposure indices based on activity, exposure duration, and distance of residence from the plant. The index of total exposure weighted for distance was associated with most respiratory symptoms and certain measures of pulmonary function. Incorporation of distance into every index increased the number of associated symptoms.

At Sambhavna, symptoms, findings and treatments are documented for every person at every visit [10]. Most patients over the age of 40 suffer from chronic obstructive airway diseases. Sixty patients were diagnosed with TB. Many patients have hypertension. Corneal opacities are found and refractive errors in young gas victims are common problems. This is the impression also at the BHMT clinics, although there is no analysis of the documentation. Menstrual abnormalities and vaginal discharge are also common symptoms. Twelve women who were in their twenties at exposure have secondary amenorrhea.

Studies on morbidity and mortality have been carried out by the Centre for Rehabilitation Studies. The morbidity pattern is still higher in the exposed cohort but there has been some decline in this pattern recently [11, 14]. The morbidity has been higher in the moderately exposed area, perhaps due to the deaths in the severely exposed area. The higher morbidity comprises mainly respiratory and gastro-intestinal problems. In 1997, the difference in the mortality rates in exposed and non-exposed communities was 4.33 per thousand [11, 15] General morbidity during the period 1996–2002 was 2–4 times

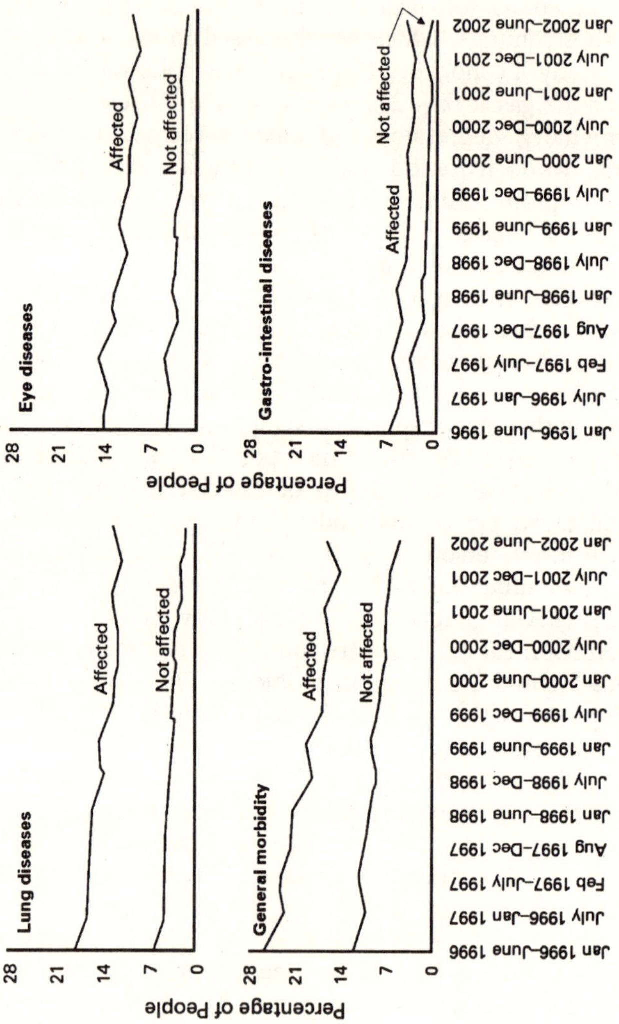

Table 5   *Morbidity in affected and non-affected areas*

higher in an exposed group compared to a non-exposed one, and gastro-intestinal diseases twice as high (table 5) [11, 16].

8.1.2.2 Eyes

The *eyes* show chronic conjunctivitis, scars on the cornea, deficiency of tear secretion or eye watering, persistent corneal opacities and the early onset of cataracts [17]. In the press conference 2003 [18], it was stated that there still is an unusually high prevalence of early age cataract. The 'posterior cortical cataract' is markedly different from the commonly found 'senile cataract'. The gas affected population is more prone to eye infections such as conjunctivitis, keratitis and scleratitis. Post operative complications are more frequent. Corneal opacity in the pupilary area of the eye is commonly found.

Eye diseases during the period 1996–2002 were twice as high in an exposed group compared to a non-exposed one (table 5) [11, 16].

8.1.2.3 Respiratory tract

Findings concerning the *respiratory tract* include abnormal lung function with obstructive and/or restrictive disease, aggravation of old diseases like tuberculosis and chronic bronchitis, and pulmonary fibrosis [17]. Sensitivity to air-borne irritants is described [19]. Permanent effects on the respiratory tract, 10 years after the leakage, were shown by IMCB [6]. Symptoms were more common for younger people. Those who lived closer to the factory had more reduced lung function.

According to an article in the *National Mail*, 30/11/1994, the doctors working in the areas affected by the gases agree that there has been a marked increase in the number of tuberculosis cases in Bhopal. In the BMTHRC annual report, 2001 [20], it is stated that there is a high incidence of tuberculosis in the population with abnormal presentation. It is further pointed out that the incidence of infectious diseases, particularly those involving the lungs, is very high. No specific study on TB has been conducted though and there is no central TB-register for Bhopal.

Lung diseases during the period 1996–2002 were 3–5 times higher in an exposed group compared to a non-exposed group (table 5) [11, 16].

8.1.2.4 Neurologic system

Neurobehavioral tests show impairment of memory, attention response speed and vigilance [17] as well as finer motor skills [7]. There are also neuromuscular symptoms such as tingling, numbness and muscular aches [17]. The investigation by IMCB [4] showed clinical signs of central, peripheral and vestibular neurological disturbance. The patients complained of headache, fatigue, difficulty in concentrating and irritation. There was no correlation with the distance to the factory.

8.1.2.5 Cytology and genetics

Some studies show impaired cell-mediated immunity in rats. The human studies, however, have limitations that make it difficult to arrive at definitive conclusions regarding the influence of the gases on the *immune system*.

Chromosomal aberrations were found to occur in exposed persons [21, 22]. An ongoing study on chromosomal changes and birth defects indicates an increased rate of birth defects in gas affected families without a previous history [18].

A population-based cancer registry has been established in Bhopal, but the onset of gas leak related cancers is not expected to occur before the 30 to 40 year lag period [17]. In the press conference 2003 [18] it was stated that there is a definite rise in the incidence of different kinds of cancers in the gas affected population over the last few years. Cancers of the lungs have increased up to 20 percent compared to other cities of the country.

Anecdotal information suggests that the number of children born with deformities such as cleft lip and missing palate is a matter of concern.

8.1.2.6 Psychological problems

The *psychological* problems are post-traumatic stress disorders (PTSD), pathological grief reactions, emotional reactions to physical problems and exacerbation of pre-existing psychiatric problems [17]. An IMCR study in 1985–1986 showed that mental health problems were as high as 132.9 per 1,000 people in affected areas compared to 24.85 per 1,000 people in non-affected areas [16]. In 1989–1990, the figure for affected areas (37.94 per 1,000) was still three times higher than the one for non-affected areas.

The impression at Sambhavna is that the patients with mental disorders, although common, do not today fulfill the criteria of PTSD.

8.1.2.7 Reproductive health

Three months after the exposure, a small study showed a high proportion of leucorrhoea, pelvic inflammatory disease, cervical erosion and/or endocervicitis, excessive menstrual bleedings and suppression of lactation [19]. Later menstrual cycle disruption, leucorrhoea and dysmenorrhoea, especially among young women were reported. Some studies, referred to by Sarangi [15], support these findings. The impressions from Sambhavna 2003 is that menstrual irregularities after the first year since menarche, profuse menstruations and premature menopause are more common among gas victims and their daughters (personal communication). No long-term studies on women's reproductive health have been done.

8.1.2.8 Endocrine system

It is expected that the released gases would have affected at least the thyroid negatively. However, no studies on this subject have been undertaken. A survey performed by Sambhavna indicates that diabetes mellitus might be more common in the gas affected population compared to a non affected one [23]. At the press conference 2003 [18] it was stated that a higher incidence of diabetes and thyroid disorders had been observed in Bhopal. However, no statistics were presented.

8.1.2.9 Children

In 1989, the still-birth rate, the crude birth rate, the perinatal death rate, the neonatal death rate and the infant mortality rate were all high in severely affected areas, compared to the rates in less affected or control areas [2].

The affected children have the same symptoms as the adults. A higher incidence of psychiatric illnesses, febrile illnesses, acute respiratory infections, gastrointestinal infections and other superficial infections of the skin, eyes and ears was reported, compared to control groups [24]. There are also reports on intellectual impairment and epilepsy [25]. Failure to grow, delay in gross motor and language sector development were found in children born a considerable time after their mothers' exposure to the gases [26].

A small study by Sambhavna [27, 28] indicates that there might be an inhibited growth among boys exposed as toddlers to the gases or born to gas exposed parents.

8.1.3 Bronchiolitis obliterans and Cor pulmonale

Breathlessness is the most predominant symptom among the survivors. The commonly used diagnosis is bronchiolitis obliterans (BO). BO is a pulmonary disease connected with many conditions, among which are toxic damage like inhalation of nitrous or chloride gases [29]. It is mainly a histological diagnosis, with thickening of membranous bronchiole including collagenous storage and fibrosis. If the lumens are constricted, the patient's air passage will be seriously and irreversibly obstructed.

The symptoms are coughing and increasing dyspnoea and obstructivity. At auscultation, a rattling sound is heard, but seldom in the rhonci. X-ray and laboratory examination is usually normal. Spirometria shows strongly reduced FEV_1 (forced expiratory volume per second) with normal or even increased diffusion capacity.

There is no cure for BO. Corticosteroids initially may reduce the inflammation. Bronchodilators seldom have a good effect on the obstructivity. The long-term prognosis for BO is poor.

When the damage to the lungs is serious, the flow of blood through the lungs is restricted. The heart must use more power, which leads to an increase in the size of the heart, the so called cor pulmonale. Too little oxygen-rich blood from the lungs reaches the body, and the patient becomes still more handicapped through restricted physical activity.

8.1.4 Post traumatic stress disorder

Post traumatic stress disorder (PTSD) was first used to describe Norwegian civilian sailors, who had been threatened by the German navy during the Second World War. Decades after the war had stopped, they were still being hit by their experiences.

Common problems are lack of control of impulses and problems with aggressiveness, susceptibility to noise, tendency to isolation, sleeping problems and tiredness [30]. Addiction problems occur to a larger extent in patients with chronic PTSD.

For a person with untreated PTSD, low degrees of stress might also activate the corticotrophine releasing factor (CRF) in the hypothalamus [31]. This leads to stimulation of most of the stress hormones: noradrenaline, adrenaline, cortisol and endorphins. American veterans from the Vietnam war had, as late as 20 years after the war, permanent and significantly more symptoms than expected from lungs, heart, muscles, joints, and the gastro-intestinal and urinary tracts, as well as severe psychological symptoms.

Re-experiencing the trauma is a necessary criterion for the diagnosis of PTSD. This may happen when the patient is reminded about the trauma, or has nightmares.

If PTSD is not treated properly, it can lead to persistent somatic symptoms [30–33].

8.1.5 Comments

The impression gained is that the information from UC as well from the Indian Government and ICMR was meant to give a picture of only mild injuries. The long-term effects were not illustrated in court. There seem to be several reasons for this fact:

• UCC wanted to reduce their liability.

- Many of the leading doctors of ICMR were closely related to UCIL.
- Would careless handling of cyanide be regarded as murder in the US?
- Multinationals did not want mother companies to be responsible for daughter companies. If this were the case, it would affect the global market.
- The Government of India did not wish to alienate multinational capital.
- There was pressure on the Government of India from the local chemical industry, which wanted no restrictions.
- The Government of India had invited chemical industries to establish themselves in India.
- The Government of India did not want to spend much money on health care.

Not one of these reasons is related to caring about the victims.

It is plausible that a significant part of the survivors' symptoms have been aggravated by untreated post traumatic stress disorder.

During the 20 years since the catastrophe, the survivors have been exposed to increasing concentrations of air pollution from smoking, cooking, diesel vehicles, two-stroke engines and the burning of waste. This has further affected their lungs and airways, as well as their eyes. Under- or malnourishment has increased susceptibility to infections. The drinking water is polluted by different microbes, causing chronic infections and malnutrition – in addition to the chemicals and heavy metals. Airway infections, including tuberculosis, are common as in all densely populated communities. Exposure to pesticides, dioxins and PVC is suspected to increase the frequency of endometriosis, with painful menstruation and infertility as results. In Bhopal, like in other third world areas, the population is ageing. Diseases like hypertension and diabetes contribute to neurological and heart symptoms.

Today, it is difficult to separate diseases and symptoms that are gas related from those that are common in any (poor) population. Only high quality epidemiological studies can prove that certain symptoms/diseases are more common among those affected by the gases. There is a desire to define the "Bhopal syndrome". This might, however, be too late.

Exposure to contaminants from the UCIL plant in food and water may affect the health of future generations.

8.1.6 References

1. Kulling, P. and H. Lorin, *The Toxic Gas Disaster in Bhopal December 2-3, 1984*. 1987, Forsvarets Forskningsanstalt (National Defence Research Institute): Stockholm, Sweden. [In Swedish]
2. *Annual Report 1990*. 1990, Indian Council of Medical Research, Bhopal Gas Disaster Research Centre, Gandhi Medical College: Bhopal.
3. *Annual Report 1992*. 1992, Indian Council of Medical Research, Bhopal Gas Disaster Research Centre, Gandhi Medical College: Bhopal.
4. Cullinan, P., S.D. Acquilla, and V.R. Dhara, *Long term morbidity in survivors of the 1984 Bhopal gas leak*. Nat Med J India, 1996. **1996(9)**: p. 5–10.
5. Callender, T., *Long-term Neurotoxicity at Bhopal*. International Perspectives in Public Health (Buffalo, NY: Ministry of Concern for Public Health), 1996. **1996(11–12)**: p. 36–41.
6. Cullinan, P., S.D. Acquilla, and V.R. Dhara, *Respiratory morbidity 10 years after the Union Carbide gas leak at Bhopal: a cross sectional survey*. BMJ, 1997. **314**(338–342).
7. Eckerman, I., *The health situation of women and children in Bhopal*. International Perspectives in Public Health (Buffalo, NY: Ministry of Concern for Public Health), 1996. **1996(11–12)**: p. 29–36.
8. Heinzow, B., *Results of the International Medical Commission on Bhopal (IMCB)*. International Perspectives in Public Health (Buffalo, NY: Ministry of Concern for Public Health), 1996. **1996(11–12)**: p. 4–8.
9. Eckerman, I., *Long-term Health Effects of Exposure to the Bhopal Gases. Observations of the hazardous effects on the health of particularly vulnerable groups*. 1995, Nordic School of Public Health: Goteborg, Sweden.
10. *The Bhopal Gas Tragedy 1984– ? A report from the Sambhavna Trust, Bhopal, India*. 1998, Bhopal People's Health and Documentation Clinic: Bhopal.
11. *Bhopal Gas Tragedy Relief and Rehabilitation Department* http://www.mp.nic.in/bgtrrdmp/setup.htm. 2004, Government of Madhya Pradesh.
12. Kamat, S., et al., *Sequential respiratory, psychologic and immunologic studies in relation to methyl isocyanate exposure over two years with model development*. Environ Health Perspect, 1992. **97**: p. 241–53.

13. Dhara, V.R., et al., *Personal Exposure and Long-Term Health Effects in Survivors of the Union Carbide Disaster at Bhopal.* Environmental Health Perspectives, 2002. **110**(51): p. 487–500.
14. Acquilla, S. and R. Dhara, *Interview with Dr. Banerjee and Moina Sharma, Center for Rehabilitation Studies.* 2004, Personal communication.
15. Sarangi, S., *The Bhopal aftermath: generations of women affected,* in *Silent Invaders. Pesticides, Livelihoods and Women's Health,* M. Jacobs and B. Dinham, Editors. 2003, Zed Books Ltd: London.
16. *Serving the symptom. The government still does not know what still afflicts people in Bhopal.* Down To Earth, 2003 (December 15): p. 29–32.
17. Dhara, V.R., *Health Effects of the Bhopal Gas Leak: A Review.* Epidemiol/Prev, 1992. **1992**(14): p. 22–31.
18. *Press Statement, Jan 30, 2003.* 2003, Sambhavna Trust: Bhopal.
19. Morehouse, W. and A. Subramaniam, *The Bhopal Tragedy. What really happened and what it means for American workers and communities at risk.* 1986, New York: The Council on International and Public Affairs.
20. *The Bhopal Memorial Hospital and Research Centre and Outreach Health Centres. Annual report 2001.* 2001, Bhopal Memorial Hospital Trust: Bhopal.
21. Ghosh, B.B., et al., *Cytogenetic studies in human populations exposed to gas leak at Bhopal, India.* Environ Health Perspect, 1990. **86**: p. 323–6.
22. Goswami, H.K., et al., *Search for chromosomal variations among gas-exposed persons in Bhopal.* Hum-Genet, 1990. **84**(2) (Jan): p. 172–6.
23. *Prevalence of diabetes mellitus, hypertension and under/over-weight among the people exposed to toxic gases from UCIL,* Sambhavna Clinic: Bhopal.
24. *The Health Problems of Bhopal Gas Victims. Assessment & Management. Working Manual 2.* 1989, Dec, Indian Council of Medical Research, DST Centre for Visceral Mechanism.: New Delhi.
25. *Clinical profile of the Bhopal gas victims. A Summary of ICMR reports.* 1994, Bhopal Group for Information and Action: Bhopal.
26. Lochan, R., *Health Damage due to Bhopal Gas Disaster. Review of medical research,* in *Compensation Disbursement.* 1992, Bhopal Group for Information and Action: Bhopal.
27. Ranjan, N., et al., *Methyl isocyanate Exposure and Growth Patterns of Adolescents in Bhopal.* JAMA, 2003. **290**(14): p. 1856–7.
28. *Study of Growth Pattern of Young People Born in Bhopal between 1982 and 1986,* Sambhavna Clinic.
29. Haeggstrom, F., A. Eklund, and G. Elmberger, *Distinguish between BO and BOOP! Two pulmonary diseases that often are mixed up.* Lakartidningen, 1997. **94**(1287–91) [In Swedish].

30. Sondergaard, H.P. *Traumatic Stress*. Lakartidningen, 1993. **90**: p. 796–800 [In Swedish].
31. Dotevall, G., *Stress and Psychosomatic Disease*. 2001, Lund, Sweden: Studentlitteratur. [In Swedish].
32. Abdulbaghi, A. and V. Sundelin Wahlsten, *Child trauma and vulnerability; new knowledge necessitates conceptual revision*. Lakartidningen, 1998. **17**: p. 1955–62 [In Swedish].
33. Koscheyev, V.S., et al., *The psychosocial aftermath of the Chernobyl disaster in an area of relatively low contamination*. Prehospital and Disaster Medicine, 1997. **12**: p. 41–6.

8.2 Epidemiological and Clinical Research

8.2.1 General aspects

Data collection has been made by several official organisations:
- The Tata Institute of Social Science (TISS)
- Indian Council of Medical Research (ICMR)
- Centre for Rehabilitation Studies (CRS)
- India Toxicology Research Centre (ITRC)
- Defence Research Development Organisation (DRDO)
- Indian Council of Agricultural Research (ICAR)
- National Environment Engine Research Institute (NEERI)
- Bhopal Cancer Register
- Council for Scientific and Industrial Research (CSIR)

It has also been done at hospitals and clinics, including Sambhavna Trust and BMTHRC.

Many independent national and international institutions, including the International Medical Commission on Bhopal (IMCB), have collected data over the years.

However, it has not always been possible to find out about what happened to the data collected. Here is a critical review of those collections from which I have been able to get some information.

8.2.2 Tata Institute of Social Science

A TISS-survey on socio-economic factors dealing with about 25,000 families was undertaken immediately after the disaster by the Tata Institute of Social Science in Mumbai. The data has still not been released.

Children born after the leakage were not included in the survey [1].

8.2.3 Indian Council of Medical Research

After the leakage, the Ministry of Gas Relief decided to set up a research institute in Bhopal for ten years. During this period, the Indian Council of Medical Research (ICMR) conducted a whole series of studies. The institute was closed down in January 1995. Up to that point, the authors were not allowed to publish their results. Only a few received a report through official channels. On the first page of the 1990 and 1991 annual reports is printed: "The contents of this report should not be reproduced, reviewed, abstracted or quoted without the written permission of the Director." Since 1995, a number of articles have been published.

Research personnel were employed on a temporary basis and several projects suffered due to the transfer of the principal investigators employed by the state government [2].

ICMR initiated 24 research projects on different areas. However, many areas like fertility and immune deficiency were not covered, although scientists suggested them. Some studies, for example, on children exposed in-utero, were terminated after six years, before definite results could be gained.

The last two projects were terminated in December 1994. There was no review of data prior to termination of the studies to determine whether some of them needed to be continued or whether any fresh studies were required to be commissioned. Members of IMCB had put forward 18 project proposals that were not taken into consideration.

The long-term effects of MIC are measured through cohort studies. From the estimated exposed population of 521,262 persons, 20.3% or 80,021 persons were chosen. The cohort has been stratified in relation to the estimated degree of exhibition, that is, in which of the areas classified as "exposed" they live. Critics consider the cohort as being rather unevenly distributed in the settlements [1]. The control group lives in an area classified as "not exposed". The method of measuring is through surveys considering symptoms. There is no follow-up of those who move outside the areas.

There is *bias* in the cohort:
- The control group might also be exposed, although to a lesser degree.
- Those who were very seriously affected by the gases, and those who have hopes of economic compensation, are more likely to remember – or even exaggerate – their symptoms (recall bias).

- Those who lived close to the factory were not only exposed to MIC, but also to many other toxic substances as well as to hypoxia.
- In the group that moved out, young women who got married predominate. This has led to underestimation of symptoms related to women's reproductive health, including malformation of babies.
- No one below 18 years was allowed to register in 1984.

There are also *confounding factors*. To live at a longer distance from the factory also means
- The composition of the gases changed;
- The risk of damage caused by hypoxia changed;
- Further away from the plant, the proportion of wealthy people increases, which in its turn means
  - less exposure because of better housing;
  - being less affected by the gases because of better health;
  - less exposure to other hazardous factors after the gas leak, such as air pollution and infections;
  - better possibilities of taking care of their own health.

Most of the clinical studies, e.g. of children and of psychological effects, were uncontrolled observations on small populations, which led to serious methodological shortcomings. Mehta et al [3] stresses that on a strictly scientific level, these studies do not conclude causality. However, many of the conclusions are supported by experimental studies.

I have been told that when a foreign scientist visited ICMR, the members suggested that he should also look at the neurological sequelae. However, this was "forgotten" by the management.

There are also other fields where the research is rudimentary or missing. Very little has been done on female reproduction, including on chromosomal aberrations.

Nothing on PTSD (post-traumatic stress disorder) was done by IMCB. This syndrome might at the time have been relatively unknown in a country like India, but should have been recognised by western expertise. In studies of "mental health", neurological and psychiatric symptoms are mixed.

Children born after the disaster are not included in the ICMR research [1].

The final report from ICMR has, 19 years after the disaster, not been released.

8.2.4 Centre for Rehabilitation Studies

The research task, including the cohort, was in 1995 given to the Centre for Rehabilitation Studies (CRS), Government of Madhya Pradesh. A fund of Rs 5 crore was made available by the Ministry of Chemicals and Fertilisers, and 50 persons have been employed on a contract basis.

Long Term Epidemiological Studies are being carried out to study the change in socio-economic and demographic pattern in the cohort through survey, to study mortalities and other vital events, to study the six monthly point prevalence and period prevalence morbidities in the cohort and to follow up on the chronically ill patients [4]. The cohort population in the affected area is 80,021, and in the non-affected area, 15,931.

The head office is in the Gandhi Medical College grounds and they also use six mini-clinics for data collection which are staffed by a doctor and a research assistant (R. A.) [5]. The R.A.s now have several years of experience and submit the questionnaires from which data is recorded and brought to the statistical cell and checked for validity. They have about a 40% attrition of sample to date in the cohort due to death, migration, etc. Migration is greater in the less exposed areas due to a better socio-economic status. The CRS is severely hampered by lack of funds, expertise, infrastructure, and bureaucratic hurdles. This prevents them from collaborating with other institutions, in the analysis and publication of their data, etc

Figures on morbidity are published on the MP web-page. However, there is no background data on how the cohort has been collected, on drop outs, on the size of the sample questioned and other facts that are necessary from a scientific point of view.

8.2.5 Bhopal Cancer Registry

The Bhopal Cancer Registry was established by ICMR under the National Cancer Registry Programme in 1986 to ascertain the magnitude of cancer problems in the central part of India and also to study the carcinogenic effects, if any, of the exposure to toxic gas at Bhopal [6]. The main objectives of the Bhopal Registry are to register all cases of cancer among the residents of Bhopal, and to generate a database to evaluate risk factors and to follow up on the population exposed to the gas.

The Government of MP has not permitted the Cancer Registry to use the cohort. In a study on cancers of the lungs, oropharynx and oral cavity, the distance to the plant at the time of the leakage was not dealt with, as it might be a confounding factor [6].

There are plans to start a cervix cancer-screening program, but so far, the CRS has not provided economic resources.

8.2.6 The Tuberculosis Hospital

As far as known, no research is done at the Tuberculosis Hospital.

8.2.7 Non-governmental Organisations

Several small studies were done by NGOs or private persons [7–12]. However, some of these studies suffer from unscientific design, bias, small sample sizes and inadequate ascertainment of exposure.

Other studies are referred to by Sarangi [13]. However, there is no list of references, so it is not possible to order the articles. The studies seem to have the same limitations mentioned here.

8.2.8 IMCB

The International Medical Commission on Bhopal (IMCB) spent three weeks in New Delhi and Bhopal in January 1994. The following subjects were studied:
• Morbidity of survivors [14–17]
• Socio-economic conditions and children's health [18]
• Compensation issues [19]

- Health infrastructure [20]
- Pharmaceutical use among survivors [21]

In the survey on morbidity, it is still unclear why the population from the "control area" had so many symptoms. Neurological examination was not performed on about a third of the chosen population, and the results of the physical examinations are not discussed in the article.

One aim was to find a way to estimate exposure dosage, using factors such as activity, exposure duration and distance of residence from the plant. The results showed that the total exposure weighted for distance has met the criteria for a successful index.

The study of socio-economic conditions and children's health took a qualitative form. The interviews were facilitated by a group of interpreters of varying quality. It gave the interviews a superficial character. The submissions might also have been biased because the women hoped to gain some kind of reward.

The compensation issue is very complicated, and the time and the experiences of the group did not allow for a high quality study.

The studies on health infrastructure and pharmaceuticals do not need comment.

8.2.9 Sambhavna

At Sambhavna, the patient records are computerised. This is used for finding data about, for example, age, sex and symptoms [2, 22]. Some small and well-designed studies on treatments have been conducted, for example, on yoga treatment [23].

Surveys on the status of health and health-care are done during home visits in the most affected areas. They are small and can only give indications [22, 24, 25]. Two studies on distribution of medicines in the gas-affected area have been completed [26, 27].

For assessment of the probable cause of death of persons in the affected area, a method called Verbal Autopsy is used, with the help of three external doctors and in co-operation with the London School of Hygiene and Tropical Medicine. The cases are sorted into different categories: related (most probable, probable, possible) and unrelated. Till today, 106 cases have been started, and 83 have been finished. Of these, 86% are regarded as related.

The presentations of the studies are sometimes not complete. Date or year could be missing.

8.2.10 Bhopal Memorial Trust Hospital and Research Centre

At the BMTHRC, as well as at the outreach clinics, all patient data are computerised. However, it is used only for simple quantitative analyses.

In the Annual report 2001 [28], several departments describe their research. However, most of this is clinical-technical research, and not specifically aimed at the gas affected patients.

At the pulmonary department, research has been initiated to look into the increased incidence of tuberculosis in the gas victims, the increased susceptibility to certain pathogenic organisms after exposure to MIC gas and the patterns of antibiotic resistance. A total of 350 houses were visited during four months in 2001. Eleven cases were picked up. However, no further progress could be made due to problems like paucity of staff, difficulty in convincing the affected population to get investigated, and problems of epidemiological nature. It is pointed out that a proper epidemiological study is required. Because of the high incidence of tuberculosis in the population, a proposed project will be submitted to ICMR.

The department of radiology in conjunction with the pulmonary department has started a study on the delayed effect on MIC gas in the pulmonary status. This might be the same study as mentioned above.

The department of microbiology is performing a comparative study of sputum, post bronchoscopy sputum and bronchoalveolar lavage in atypically presenting pulmonary tuberculosis in gas victims.

The nephrology department has initiated a project to estimate the burden of chronic renal disease in a large population based project. Nothing has been said about comparing the gas affected to the non gas affected.

For the outreach clinics, no research, studies or monitoring are described.

8.2.11 Comments

Although the quality of the clinical research varies, the different reports support each other. The findings are also supported by animal experiments.

We are still waiting for the final report from ICMR. We are also waiting for ICMR to release the names, addresses, etc., of the cohort, so that the Centre for Rehabilitation Studies and the Bhopal Cancer Registry can continue following the cohort.

The official set-up for monitoring exposure-related deaths was disbanded in December 1992 [2, 29]. This means that late cases caused by respiratory and/or cardiac insufficiency, cancer and TB will never be highlighted.

In the procedure involved in getting their final compensation, the survivors must hand over their papers. In the future, it may be very difficult to do "sound epidemiology" on exposure and disease.

A programme for "outbreak disaster epidemiology" should be drawn up. The WHO could be responsible for this.

The recommendations could include the following parameters:

- Find all the important parameters for registration.
- Register the entire population, including children and the non-affected.
- Choose a cohort, of all ages, for long time studies. Every person must be followed up, even if they move.
- Choose certain groups like children, or fertile women, out of the cohort for more detailed studies.

To improve studies of the disaster that led to exposure to chemical and/or radioactive compounds:

- Include monitoring for cancer, reproductive health, hormone systems and neurological systems. Always include monitoring

for psychological symptoms (post-traumatic stress disorder – PTSD). If several organisations are collecting data, there should be coordination among them, and it should be possible to combine the different databases.
- A combination of panel, cohort and case – control designs may be used to provide a more detailed description of the range of health effects experienced by the population [17].

8.2.12 References

1. *Compensation Disbursement. Problems and Possiblities.* 1992, Bhopal Group for Information and Action: Bhopal.
2. *The Bhopal Gas Tragedy 1984– ? A report from the Sambhavna Trust, Bhopal, India.* 1998, Bhopal People's Health and Documentation Clinic: Bhopal.
3. Mehta, P.S. and E. AL, *Bhopal Tragedy's Health Effects. A review of methyl isocyanate toxicity.* JAMA, 1990. **264**: p. 2781–7.
4. *Bhopal Gas Tragedy Relief and Rehabilitation Department* http://www.mp.nic.in/bgtrrdmp/setup.htm. 2004, Government of Madhya Pradesh.
5. Acquilla, S. and R. Dhara, *Interview with Dr. Banerjee and Moina Sharma, Center for Rehabilitation Studies.* 2004, Personal communication.
6. Dikshit, R., *Lung, Oropharynx and Oral Cavity Cancer in Bhopal.* 1998, University of Tampere: Tampere, Finland.
7. Eckerman, I., *Long-term Health Effects of Exposure to the Bhopal Gases. Observations of the hazardous effects on the health of particularly vulnerable groups.* 1995, Nordic School of Public Health: Goteborg, Sweden.
8. Kapoor, R., *Fetal loss and contraceptive acceptance among the Bhopal gas victims.* SocBiol, 1991. **38**(3–4): p. 242–8.
9. Satyamala, C., N. Vohra, and K. Satish, *Against all Odds. Continuing Effects of the Toxic Gases on the Health Status of the Surviving Population in Bhopal.* 1989, CEC: New Delhi.
10. Sathyamala, C., *Fertility and Gynaecological Disorders: Impact of Bhopal gas leak disaster,* in *Dep of Epidemiology and Population Sciences.* 1993, London School of Hygiene and Tropical Medicine: London.
11. Sathyamala, C. and e. al., *Effect of Bhopal Gas Leak on Women's Reproductive Health.* 1985, IBCS: Bombay.
12. *Distorted Lives. Women's Reproductive Health and Bhopal Disaster.* 1990, Medico Friend Circle: Pune.
13. Sarangi, S., *The Bhopal aftermath: generations of women affected,* in *Silent Invaders. Pesticides, Livelihoods and Women's health,* M. Jacobs and B. Dinham, Editors. 2003, Zed Books Ltd: London.

14. Cullinan, P., S.D. Acquilla, and V.R. Dhara, *Long-term morbidity in survivors of the 1984 Bhopal gas leak.* Nat Med J India, 1996. **1996**(9): p. 5–10.

15. Cullinan, P., S.D. Acquilla, and V.R. Dhara, *Respiratory morbidity 10 years after the Union Carbide gas leak at Bhopal: a cross sectional survey.* BMJ, 1997. **314**(338–342).

16. Callender, T., *Long-term neurotoxicity at Bhopal.* International Perspectives in Public Health (Buffalo, NY: Ministry of Concern for Public Health), 1996. **1996**(11–12): p. 36–41.

17. Dhara, V.R., et al., *Personal Exposure and Long-term Health Effects in Survivors of the Union Carbide Disaster at Bhopal.* Environmental Health Perspectives, 2002. **110**(51): p. 487–500.

18. Eckerman, I., *The health situation of women and children in Bhopal.* International Perspectives in Public Health (Buffalo, NY: Ministry of Concern for Public Health), 1996. **1996**(11–12): p. 29–36.

19. Jaskowski, J., et al., *Compensation for the Bhopal Disaster.* Int Perspectives in Public Health (Buffalo, NY: Ministry of Concern for Public Health), 1996. **1996**(11–12): p. 23–28.

20. Verweij, M., S.C. Mohapatra, and R. Bhatia, *Health Infrastructure for the Bhopal Gas Victims.* Int Perspectives in Public Health (Buffalo, NY: Ministry of Concern for Public Health), 1996. **1996**(11–12): p. 8–13.

21. Bhatia, R. and G. Tognoni, *Pharmaceutical use in the victims of the carbide gas exposure.* International Perspectives in Public Health (Buffalo, NY: Ministry of Concern for Public Health), 1996. **1996**(11–12): p. 14–22.

22. *Prevalence of Diabetes Mellitus, Hypertension and Under/Over-weight Among the People Exposed to Toxic Gases from UCIL*, Sambhavna Clinic: Bhopal.

23. Gupta, A. and S. Durgvanshi. *Yoga therapy in the care of chronic respiratory disorders among survivors of the December 1984 Union Carbide disaster in Bhopal.* in *The 3rd International Conference on "Yoga Research & Traditions".* 1999, Jan 1–4. Kaivalyadhjama, Lonavia.

24. *Study of Growth Pattern of Young People Born in Bhopal between 1982 and 1986*, Sambhavna Clinic.

25. Ranjan, N., et al., *Methyl isocyanate Exposure and Growth Patterns of Adolescents in Bhopal.* JAMA, 2003. **290**(14): p. 1856–7.

26. *Assessment of treatment offered at the Bhopal Hospital Trust's community clinic no.1 and analysis of Bhopal Hospital Trust prescription data.* 1998, Sambhavna Clinic: Bhopal.

27. *Report of Survey for Assessment of Drug Distribution in Gas Affected Bhopal*, Sambhavna Clinic: Bhopal.

28. *The Bhopal Memorial Hospital and Research Centre and Outreach Health Centres. Annual report 2001.* 2001, Bhopal Memorial Hospital Trust: Bhopal.

29. *13th Anniversary Fact Sheet on the Union Carbide Disaster in Bhopal*, in *Compensation Disbursement*. 1997, Bhopal Group for Information and Action: Bhopal.

8.3 Social and Economic Rehabilitation

8.3.1 The bureaucratic organisations

After the accident, no one under the age of 18 was registered as a victim. Yet, the number of children exposed to the gases is estimated to be around 200,000, while 1,000–4,000 children died within the first weeks. It is not known how many children were orphaned.

The Central Government of India decided in 1984 to establish a separate vertical structure for the relief of the gas victims. The Ministry of Gas Relief is still in operation, under the Central Government's Ministry of Fertilisers and Chemicals. Under the Chief Minister in Madhya Pradesh, the Department of Gas Relief and Rehabilitation was created. Responsibility for the funds for rehabilitation of survivors was meant to be shared between these two departments.

The first action plan for the medical and economic rehabilitation of the victims was to the tune of Rs 188.50 crore and should have been completed by 1995. In 1998, the government sanctioned an additional amount of Rs 69.50 crore for completing the incomplete work of the first action plan (hospital construction and furnishing, economic rehabilitation, provision of drinking water and social rehabilitation). It made it clear that the funds for the second action plan would be released only if all the work of the first action plan was completed by September 30, 1998 [1]. In June 2000, the Union Government announced the institution of a high level investigation, alleging misuse of funds [2]. Due to lack of funds, provision of health care is almost the only activity of the Department.

The second action plan involves Rs 319 crore. Of the budgeted Rs 245 crore (US$ 70 million), 51 percent is meant for furnishing and running the Kamla Nehru Hospital, and a total of 4 percent is allocated for community health schemes, service oriented research and an indigenous system of medical care [3].

A 60-member State Advisory Committee was formed in 1987 to "prevent over-governmentalisation of the Bhopal Relief and Rehabilitation Programme by involving non-officials in the decision making process" [4]. The committee holds one or two meetings per year, and very few of its decisions are ever implemented.

In December 1994, the Government of Madhya Pradesh advertised in the daily papers for NGOs and other social or professional organisations that are interested in planning and implementing small enterprises as well as running welfare programmes. In April 1999, there was still no sign of these programmes.

8.3.2 Immediate relief

Two days after the tragedy, the Government fixed the following compensation to be given to victims [5]:

a. Rs 10,000 to the relations of the deceased;
b. Rs 1,000 and less for those not seriously affected;
c. Rs 1,500 to each family in the 36 wards, reporting a monthly income of less than 500 Rs.

Relief measures commenced in 1985, during which milk, bread, sugar, and edible oil were distributed. Subsequently, this distribution was discontinued and the victims were neglected [6].

Before the elections took place, the local government made free rations available to the majority of the population. About 700,000 ration cards were issued. In January 1985, following the elections, the government halted the free distribution of rations, to investigate who was entitled to get relief. After large-scale protests from the residents, the rations were restored [7].

A number of children were orphaned. According to Chauhan, [6], 31 orphaned children and six destitute women were picked up and kept at a centre specially opened for them. Of these children, 25 were later reunited with their relatives. The State Government allotted a plot of land for the establishment of an SOS village. The MP Government [8] describes that "27 orphans were maintained at Government expense till they became adult. Till 31.3.2003, three such orphans are maintained by various NGOs at Government expenses."

Widow pension at the rate of Rs 200/- p.m. was provided to widows of the deceased gas victims till the final award of compensation was made by the Court of the Welfare Commissioner [8].

8.3.3 Economic compensation

8.3.3.1 Compensation from Union Carbide

The process of claims is described by Jaising [9], Jones [7], Jaskowski et al [10], Cassels [11], BGIA [12], Pandey [13], Morehouse [14], Lapierre and Moro[15], and in the Sambhavna report 1998 [16].

It was in the interests of UCC not to recognise the serious effects of the gases. The amount of compensation the company would ultimately be liable to pay would depend on the extent of the injuries. In addition, the Government of India had – and still has – an interest of keeping the compensation claim low. A high claim would discourage other chemical industries from establishing plants in India.

Just a few days after the disaster, American lawyers invaded Bhopal and persuaded the victims to put their thumbprints on powers of attorney. In the US, they then claimed billions of dollars as compensation.

In March 1985, the Government of India passed the Bhopal Gas Leak Disaster Act. It gave the government the statutory right to represent all victims in or outside India. The government constituted itself as the sole representative of the victims, with full authority to litigate on their behalf and to settle their claims. The Act empowered the government to frame a scheme for registering and scrutinising the claims of the victims.

This Act almost completely disempowered the victims from participating in the adjudication process. It denied the victims any form of control over their own lives. For Union Carbide, this act was to great advantage. A settlement with the Government of India would have the authority of a statute and would not need to be justified as fair in a court of law [9].

In 1985, the Union of India filed a plaint before District Court, New York [13]. After arguments from UCC, Judge

Keenan ordered that UCC should submit to the jurisdiction of the Courts of India, and that temporary relief should be provided. In 1986, the Government of India filed a suit against UCC at the District Court of Bhopal. UCC was restrained from selling its assets while the suit was pending, and was ordered to pay Rs 350 crore as interim relief. This sum was reduced to 250 crore by the High Court of Madhya Pradesh. In 1988, UCC and the Union of India filed protests with the Supreme Court.

In 1985, Union Carbide made an offer of US$ 350 million, the insurance sum. The Union of India refused to accept the offer most emphatically and claimed US$ 3.3 billion. In February 1989, the proceedings were still underway in the Supreme Court of India. There were few signs of progress, as Union Carbide denied all liability. After lunch on February 14, it was announced that a settlement had been arrived at. Union Carbide agreed to pay US$ 470 million in full and final settlement of its civil and criminal liability.

The sum US$ 470 million corresponds to the insurance sum of US$ 350 million plus the interest [9]. According to the report from Sambhavna [16], the amount paid as compensation is Rs 715 crore. The annual interest would be more than Rs 100 crore. Around Rs 1,300 crore remains to be disbursed*.

The court categorised the compensation of Rs 750 crore as follows [5]:
a. Rs 70 crore for 3,000 fatal cases (range of compensation Rs 1 lakh to Rs 3 lakh each).
b. Rs 80 crore for 2,000 cases of severe injury (Rs 4 lakh each).
c. Rs 280 crore for 30,000 permanent total or partial disability cases (Rs 50,000 to Rs 2 lakh each); for 20,000 temporary, total or partial disability cases, Rs 100 crore (Rs 25,000 to Rs 1 lakh each) for creation of special institutional facilities for medical treatment; Rs 25 crore for 50,000 cases with minor injuries – Rs 100 crore (Rs 20,000 each) for 15,000 cases with loss of livestock etc – Rs 50 crore (Rs 10,000 each).

---

* On July 19, 2004, the Supreme Court ordered the government to distribute the remaining $325.5 million (Rs 1,503 crore) to the victims and relatives of the dead. The unspent money grew over the years because of the interest and the rupee's depreciation.

The Supreme Court decision has been challenged and a number of review and writ petitions have been filed against it before the Supreme Court [5]. Some important points raised in the petitions were as follows:

1. The Bhopal Gas Leak Disaster (Processing of Claims) Act of 1985 has itself been challenged as unconstitutional since it takes away the fundamental right of the citizen to choose their own representatives. A decision on the same is still awaited. Neither the court nor the government could take a decision which affects the parties involved in the case.
2. The decision terminates all Civil and Criminal proceedings which may not be just.
3. The persons who are the beneficiaries and who would get compensation were not given an opportunity to be heard.
4. The amount of compensation has been arrived at rather arbitrarily and is grossly inadequate. The main arguments in support of this contention have been
   a) Deduction of the amount of USD 5 m donated to Red Cross was not justified since no conditions were imposed while making the donation.
   b) As referred to earlier, the Government of MP had worked out 15 categories of claims. The decision, however, was based only on eight of them. There is no provision for the remaining seven categories of claims.
   c) The original claims were for USD 3,000 m.
   d) The figures of victims under each category, which form the basis of the decisions, have been called into question.

In 1985, the American judge Keenan decided that UCC should pay 5 million US dollars to the American Red Cross Society to be transferred to the Indian Red Cross Society, and four community clinics were set up. Four years later, UCC had the Indian Red Cross pay back all the remaining funds, and the clinics were eventually closed down.

The October 3, 1991 judgement of the Supreme Court has made provisions for "later born children who might manifest congenital or prenatal MIC related afflictions" [11].

In 1991, the Supreme Court of India directed UCC to finance a 500-bed hospital for medical care for the survivors. When

UCC continued to be absent in court, the judge of the Bhopal District Court directed the attachment of UCC's shares in UCIL. Within three weeks of the warning, UCC registered a "charitable trust" in London, with a sole trustee. UCC put £31,000 in the trust and for the rest of the trust's assets, pledged its shares in UCIL. UCC appeared to have used its influence in the Supreme Court, and was allowed to sell the attached shares. Big sums were used for the benefit of the trustee [17].

In March 1999, the Government had not yet released the Rs 270 million sanctioned during implementation of the first action plan.

8.3.3.2 Disbursement of compensation to the survivors

No one under 18 years of age was allowed to register as a victim, and government functionaries stated that minors were not entitled to compensation [9]. Later, they might be recognised as victims if the family had a ration card. However, in 1991, a survey carried out by an NGO showed that 20 percent of the people in the area did not have ration cards [9]. Figures obtained from the Directorate of Claims indicate that the number of claimants in the 26 wards was 16 percent less than the official resident population in 1984 and 33 percent less than the population of the 36 wards in 1989 [12]. There has been no governmental effort to monitor the migration that has taken place in and out of the gas-affected area.

The State Government decided that all families in the gas-affected area, whose monthly income was Rs 500 or less, should be given a one-time ex-gratia payment of Rs 1,500. More than 70,000 families received this compensation [18].

Under the Social Security Pension Scheme, from November 1985, 302 widows were granted a pension of Rs 300 per month, later raised to Rs 750. A pension of Rs 60 per month was given to 1,042 destitute women and 440 men [18].

Lost livestock was compensated for at the rate of Rs 3,500 for a buffalo on a descending scale to Rs 20 per head for poultry. Compensation was paid for a total of 2,135 dead animals, at a cost of Rs 2.3 million. Nearly 28,000 animals were treated by veterinary doctors.

In 1987, the ICMR drew up rules for categorisation (blood and urine tests, x-rays etc.). Each claimant was to be categorised by a doctor. In court, the claimants were expected to prove "beyond reasonable doubt" that death or injury in each case was attributable to exposure. Victims, who did not have medical records or the money to pay the doctor, were said to have had great difficulties in getting into the right category. It is also documented that the officers made decisions on their own, and that they seemed to keep some money for themselves [17]. Most victims had no bank accounts and were required to open accounts after depositing 20 rupees before the banks would handle their cheques.

The criteria for the medical categorisation were regarded as inadequate. In 1992, 44 percent of the claimants still had to be medically examined. Of the categorised claimants, 92 percent had been categorised as having suffered only temporary injuries or no injuries at all [12].

The interim relief, which had been paid to the inhabitants of 36 wards since 1990, was Rs 200 (around US$ 7) per month per person and was paid to everyone in the family who was born

Court (Christian Saltas)

before the accident. In 1994, there was a proposal to provide interim relief of Rs 200 per month to the inhabitants of all 56 wards and to consider two wards as severely affected, five wards as highly affected and 47 wards as generally affected.

In 1993, decisions about final compensation were speeded up. One court was opened in each of the original 36 wards. The costs for personnel and houses were deducted from the sum available for compensation.

To get the final compensation, six documents were necessary. Medical prescriptions and reports were also asked for. Everything except the ration card, which was the common property of the whole family, was kept by the court. The victims had to sign a paper saying that they agreed and that they would make no further claims before the judgement. Most people did sign, because they had already waited for more than ten years.

According to the Supreme Court directive, those directly affected should be provided with the interim relief ranging from Rs 25,000 to Rs 400,000. The SC directive concerning death cases assured Rs 100,000 to Rs 500,000 [19]. The final compensation for personal injury is for the majority Rs 25,000 (US$ 830). This sum is reduced by the interim relief paid out earlier, around Rs 10,000. Those who have received interim relief for six years will thus get only Rs 10,500 (US$ 350) as the final compensation. This compensation will be paid to all claimants [20]. For death claims, the average sum paid out is Rs 62,000.

The size of the compensation was set in 1989. Because of inflation, the sum of Rs 25,000 has lost half of its value since 1989.

The compensation sum of US$ 470 million was estimated on the basis of 3,000 cases of death and 200,000 cases of injury. The population of the affected areas in 1984 has been estimated to be 520,000 in the 36 wards. In 1988, it was estimated as 565,000. In 1995, more than 618,000 cases in the 56 wards had been filed: 597,300 personal injuries, 15,200 deaths and 5,000 others. The approximate number of gas victims who failed to file their claims was estimated at 100,000 [18]. According to official records, in 1999, 1,001,723 claims for injury and 22,149 claims for deaths were registered in Bhopal [21], i.e officially, local records show that there were about five times more claims than the US compensation calculations.

The official body for registration of exposure related deaths was wound up in December 1992.

In 1997, about 30 percent of injury claims had been rejected, and 400,000 claims waited to be decided. Of the death claims, 3,900 (25 percent) were rejected, and 11,300 cases were paid compensation [1].

In 1999, the proposal that all 56 wards should be considered to be gas affected engaged the survivors' organisations, which also held different opinions on this issue. In a letter to the Chief Minister of Madhya Pradesh, three of the most active organisations pointed out that in the documentation compiled, there is no evidence that residents outside the 36 areas are likely to have been injured by the gases. If the compensation is also disbursed in the non-affected areas, it would mean less money for the residents in the affected areas. Furthermore, the residents in the non-affected areas are relatively wealthy, and the money would be less important to them.

In 2000, nearly all the "old" cases were decided upon [2]:

	Personal injury cases		Death cases*	
	Old	New	Old	New
Registered claims	597,908	403,815	15,310	6,839
Decided cases	597,908	395,421	15,308	6,271
Awarded cases	541,809	188,108	11,628	2,782
Rejected cases	56,099	207,313	3,680	3,489
Pending cases	—	8,394	2	568
Total compensation amount	920.61 crores	470.38 crores	77.21 crores	7.32 crores
Average compensation	26,665	25,006	66,400	26,312

* Over 7,000 deaths have been judged to be exposure-related with the majority of the death claims being "converted" arbitrarily into injury claims.

In 2000, the Welfare Commissioner issued a notification that claimants whose claims have been rejected for reasons of "non-appearance" will be denied their right to compensation. The survivors' organisations protested strongly, as many of the "non-appearing" claimants had not received any notification at all. It was admitted that 40,000 claim files were missing from the Directorate of Claims, and some 20,000 claims had been taken by impostors.

Union Carbide's defence lawyers argued that an American court was not competent to assess the value of a human life in the Third World. "How can one determine the damage inflicted on people who live in shacks?" One newspaper did the arithmetic [15]. An American life was worth 500,000 dollars. India's gross national product was 1.7 % of USA's. If the Court were to provide compensation for each deceased Indian victim proportionately, that would give 8,500 US$ (around Rs 400,000) per victim.

Because of the smallness of the sums paid and the denial of interest to the claimants, a sum as large as Rs 1,000 crore is expected to be left over after all claims have been disposed of. This has prompted the ruling party (BJP) to raise the demand of paying compensation to residents of the city living in posh areas who have no history of gas exposure [21]. Survivors and solidarity organisations demand that the balance of the compensation funds should go to a "National Commission on Bhopal", constituted of professionals, representatives of survivors' organisations as well as government officials.

8.3.4 Occupational rehabilitation

In 1985, workers and survivors demanded that the plant be reopened [22]. Plans were made to set up cooperatives to manufacture milk- and soyaproducts, textiles, etc.

The Special Training and Employment Programme for the Urban Poor (STEP-UP) was an ongoing scheme that was used for the gas victims [18]. The strategy included the upgrading of skills in appropriate trades, like sewing, stitching and production of stationery items, combined with the grant of a loan of up to Rs 12,000 for self-employment. Fifty worksheds were to be constructed, and 2,500 persons were to be trained. Till December 3, 2003, about 28,700 beneficiaries have been provided loans to the tune of Rs 19.20 crore under this programme for self employment [8].

Only 33 of the worksheds were ever started. The number of women employed was 2,300. The worksheds ran at a yearly profit of Rs 1 crore [17]. They were also a meeting place for the women. All except the Women's Stationery Workshop were closed down by 1992 [23]. Their fight for survival has already been described [23]. As late as 2000, they were not always paid salaries. The same year, a worker lost his thumb in one of the paper-cutting machines, as the safety system was missing [2].

In 1986, the MP government invested Rs 8 million (US$ 2.3 million) in the Special Industrial Area Bhopal [1, 23]. Here, 170–200 worksheds and common facilities like a tool room, postal services, a dispensary and a communication centre were planned. Employment was meant to be provided to 10,000 persons. Vocational training for 3,600 candidates per year was planned. Only 152 worksheds were built [8]. After some years, parts were rented to private enterprises that promised to employ gas victims. Other parts of the buildings are being used by the home guard. Of the worksheds, 16 were partially functioning in 2000 [2]. The figures of how many persons received training here varies from 6,461 [21] to 2,443 and 461 [2]. It is not known how many of these were gas victims.

In 1992, 200 gas-affected women were selected for a three-month training period in the production of jute handicrafts.

An Industrial Training Institute that started in 1994 still remained fully operational [21] in 1999. Till March 31, 2003, about 8000 trainees have been benefited [8].

According to the web-page of the Government of Madhya Pradesh, 42 worksheds have been constructed in gas affected areas which have been handed over to various Non-Governmental Organizations who, in turn, provide employment to gas victims by running various training cum production centres [8]. From the lists following this text, it seems that 27 such centres exist today.

In 1994, 13 local NGOs, most of which were connected to the then ruling party, were entrusted with running training-cum-production centres. By 1998–1999 only two of these NGOs were running any employment generation program. In 1998, the production of jute handicrafts was begun and offered employment to 400 women. However, in 1999, this programme was terminated on the grounds that there was no market for the goods produced [21].

The Bhopal Gas Peedit Mahila Udyog Sanghatan (BGPMUS) has set up a training-cum-production centre to be run by the Swabhiman Mahila Prashikshan Sanstha [2]. In 2000, it employed 1,000 women who prepared jute bags and did stitching work.

The Women's Stationery Trade Union has made a proposal on starting a co-operative for training and production of office production.

The MP Government has announced plans to start colleges of medicine and engineering for the rehabilitation of gas victims. It is not known how many of the young gas victims have the necessary education to be allowed to enter these colleges.

Union Carbide runs a rehabilitation programme for their employees, which gives jobs to six to ten people. They are put in different workplaces, so that they cannot communicate with each other.

It is estimated that 50,000 persons need alternative jobs, and that less than 100 gas victims have found regular employment under the government's scheme [17].

The Comptroller and Auditor-General of India (CAG) has pointed out that only 4,080 were trained in fewer than 25 trades, compared to the target of 3,600 persons each year in 40 different trades between 1990 and 1999 [2].

Twelve schools (5 primary and 7 middle schools) have been constructed by the government in the gas affected areas to provide easy accessibility to the children of the gas victims to schools, so as to educate them and bring them back into the social mainstream [8]. Apart from these, six schools in the gas affected areas have been repaired or have been upgraded under this scheme. In addition, 19 buildings of the Bhopal Gas Tragedy Relief and Rehabilitation Department have also been handed over to the local municipal authority for their use for the schools.

8.3.5 Habitation rehabilitation

With a view to providing houses (free of cost) to the kith and kin of the deceased gas victims, 2,486 houses were constructed and allotted [8]. These two- and four storey buildings with very small one room flats have been constructed outside the town ("Widows colony"). The widows do not pay any rent.

The height of the houses is a problem to those supposed to live there, as many have difficulty in walking up a set of stairs because of respiratory dysfunction. The water from the water tower does not reach the third and fourth floors; it has to be carried up. It is not possible for the families to keep cattle. There is still infrastructure missing, like buses, schools, banks, workplaces and health centres.

The problems were described in an article in *Hindustan Times*, 2001 [24]: 1,500 families lived "amidst foul smell, choked drains, dry taps and water supply lines running into sewage lines". Water had to be fetched one kilometre away. In 1998, eight people died of cholera. Those who found better alternatives have left the colony, deserting more than 1,000 houses. When the gas widows could not pay the electricity bills, the connections were cut and the meters taken away. Most of the houses are now stealing electricity from the main line. The widows are also expected to pay property tax.

In 1991, during the rainy season, 756 houses in a mildly exposed area, all but four of which belonged to Muslims, were demolished with bulldozers under the close supervision of an armed police force [11]. The families were taken out of town to form new settlements, provided with a small amount of money, some building materials, no roads but a few water taps. It is said that the new authorities used part of the moneys meant for the victims for this "beautification programme" [15].

Housing and environmental hygiene conditions have improved greatly since 1984 due to improvement in socio-economic status [25]. Today, the houses opposite the UCIL plant are "pucca" houses, compared to the "kuccha" houses of 1984. The standard of living has risen since 1984 [23]. There is evidence that the economic compensation combined with permanent employment has made this possible [5].

8.3.6 Environmental rehabilitation

In 1985–1986, the factory was closed down. Pipes, drums and tanks from the production union were cleaned with water and a chemical decontaminant and then sold off to local entrepreneurs [26]. However, the MIC plant, the Sevin plant, the tanks and the control room are still there, as are storages of different residues. Isolation material is falling down and spreading all over the place. The grass is growing high. The fence is broken, the guards seem to have disappeared. In 1997, the MP Government's minister for culture suggested turning the site into an amusement park. The protests caused the authorities to withdraw the proposal. The victims' organisations fight for a memorial at the site [21].

The area around the plant was used as a dumping ground area for hazardous chemicals. Between 1969 and 1977, all effluents were dumped in an open pit near the eastern wall of the factory [27]. From then on, neutralisation with hydrochloric acid was undertaken in two lime pits, from where the effluents went to two evaporation ponds, behind and outside the factory. The evaporation ponds were lined with a film of polythene to prevent seepage. In the rainy seasons, the effluents used to

*New **bastis** have grown up on the contaminated ground outside the UCIL plant (Ingrid Eckerman)*

overflow and enter the sewage that passed through J.P. Nagar, a slum cluster opposite the main gate of the factory.

It was in 1982 itself that tubewells in the vicinity of the UCC factory had to be abandoned [28].

Carbide's own laboratory tests in 1989 revealed that soil and water samples collected from near the factory were toxic to fish [29]. The report noted that 21 sites inside the plant were highly polluted. Samples drawn in June-July '89 from land-fill areas and effluent treatment pits inside the plant were analysed [30, 31]. They consisted of nine soil/solid samples and eight liquid samples. The solid samples had organic contamination varying from 10% to 100% and contained known ingredients like napthol and naphthalene in substantial quantities. The majority of the liquid samples contained napthol and/or Sevin in quantities far higher than permitted by ISI for onland disposal. All samples caused 100% mortality to fish in toxicity assessment studies and were to be diluted severalfold to render them suitable for survival of fish.

Analysis undertaken in 1990 at the Citizens Environmental Laboratory in Boston, showed the presence of toxic chemicals

Pump with contaminated water (Christian Saltas)

in the surroundings of the plant [2]. The same year, the National Environmental Engineering Research Institute (NEERI), an Indian government research agency, published a report that said there was no significant contamination.

In 1991, the State Research Laboratory of the Public Health Engineering Department reported serious chemical contamination in samples taken from 11 tubewells in the area. This laboratory repeated its exercise in 1996 and reported similar results [28]. The municipal authorities declared water from over 100 tubewells to be unfit for drinking, but did nothing to provide safe drinking water to the affected communities [16].

In 1994, a collaborative investigation was done with a UCC paid company and an Indian government agent. The study, that is still confidential, reports that 21% of the factory premises are seriously contaminated with chemicals, and recommends a fuller investigation for better assessment of the environmental

Drinking contaminated water (Christian Saltas)

contamination [17]. A total of 44 tonnes of tarry residue and 2.5 tonnes of alpha napthol were found and shifted to a covered area [29].

In 1996, the Central Bureau of Investigation (CBI) investigated the plant. It directed an expert body from the Madhya Pradesh Pollution Control Board, NEERI and HCT clean up the plant premises [29].

In 1999, samples of solid wastes, soils and groundwater from within and surrounding the site were collected by Greenpeace International [32]. Samples collected within the plant area contained elevated levels of mercury and organochlorine compounds. Volatile organochlorine compounds (VOCs) were found in groundwater and wells. Samples from the former solar evaporation ponds appeared to be less contaminated. It might be related to the company management digging up the bottom soil from the solar evaporation ponds and burying the sludge under three metres of farm soil in 1996.

In 1999–2002, soil samples, groundwater and vegetables from the residential areas around UCIL and from the UCIL factory area showed contamination with heavy metals such as mercury, chromium, copper, nickel and lead, as well as toxic organochlorines, hexachloroethane, hexahlorobuta-diene, pesticide HCH (BHC), volatile organic compounds and halo-organics [27, 29]. The same contaminants were found in breast milk.

In 2001, groundwater was analysed for mercury [33]. The concentrations were highest in the samples taken from sites north of the factory. Elevated concentrations were detected as far as 2.5 km northwest of the plant.

Under the original land lease agreement, Carbide was supposed to clean up the factory site before giving it back to the government [34]. It violated that agreement by abandoning the factory site after the disaster with thousands of tonnes of toxic waste still lying around. It is also said that lots of chemicals are buried in the grounds.

In order to provide safe drinking water to the gas victims who are residing in and around the Union Carbide factory, a scheme of Rs 3.00 crore has been sanctioned covering ten gas affected wards [8]. The scheme is being executed through the local municipal authority for which a sum of Rs 2.53 crore has already been released to the local municipal authorities.

In 2002, there were many new water towers spread out over the town. The water in the lakes is continuously going down. The plan is to lead in water from one of the nearby rivers. But so far, it has not been decided from which river to fetch the water. It is not clear whether this water also will reach the areas with contaminated wells.

The Government of Madhya Pradesh has compiled a list of rehabilitation measures [8, 20] that includes pumps and tree planting. But money meant for improving the living conditions of the survivors was spent on routine municipal activities like resurfacing of roads, planting of trees and construction of drains, in areas that were not or only slightly affected by the gases. Many of the pumps are said to be dysfunctional after roadworks, and the water is said to be contaminated from the sewage drains. Distribution of smokeless chulah with 50 percent subsidy has also been undertaken.

8.3.7 Comments

With the Bhopal Gas Leak Disaster Act, the Government of India became the "lawyer" of the victims, as well as part owner of the company. Today, the victims are fighting their "lawyer" to get their rights.

The survivors, who after the leakage have become even poorer than they were before, must find money for bribes and fees to get their economic compensation.

A dependable system for economic interim relief ought to have been developed much earlier. Directly after the leakage, every person, also children and the deceased, should have been registered in a reliable way. This register should have become the basis for compensation and rehabilitation.

A system of life-long pensions for those who cannot support themselves should have been developed, to assist groups like widows, orphans, the chronically ill and disabled survivors. It is pointed out that often in India, there is a long interval between approving a government loan or a grant and its actual payment. Dr Shrivastava concludes, "The amount of money being spent on relief operations is not necessarily going either to the victims or for needed work" and "the more vocal, aggressive and politically well connected people are getting relief benefits quicker and in larger quantities than some of the more needy but powerless victims" [35].

Today, the fight for environmental rehabilitation is strong. Dumping chemical waste in nature was the general way of getting rid of it in all countries during the 19th and 20th centuries. UC was not worse than anyone else. What differs is that, thanks to the activists, the polluting consequences are visible and have been made public. There is no reason why the rule "polluter pays" should not be used in this case.

Dow Chemicals disputes all responsibilities for the victims and today's situation. This should be compared to ABB, the Swedish multinational that took over the American company Combustion Engineering around 1991. This company had until the 70s used asbestos as insulation material. ABB took over the compensation claims for those who fell ill. In 2001, it became possible also for persons who had not yet any symptoms to claim compensation. ABB has offered to pay more than 1 milliard dollars in compensation. (*Dagens Nyheter*, Jan 18, 2003.)

8.3.8 References

1. 1998, Oct 2., The Indian Express. Website.
2. *Union Carbide Disaster in Bhopal. Fact sheet. 16th Anniversay of the December 2–3, 1984.* 2000, Bhopal Gas Pidit Mahila Udyog Sangathan and Bhopal Group for Information and Action: Bhopal.
3. *13th Anniversary Fact Sheet on the Union Carbide Disaster in Bhopal*, in *Compensation Disbursement*. 1997, Bhopal Group for Information and Action: Bhopal.
4. *Stop the Ongoing Medical Disaster. Bhopal Healthcare is sick.* 2000, Sambhavna Trust: Bhopal.
5. *Socio-economic Impact of Disbursement of Interim Relief to Gas-affected Families of Bhopal.* 1991, Academy of Administration, Government of Madhya Pradesh: Bhopal. p. 154.
6. Chauhan, P.S., *Bhopal Tragedy. Socio-legal Implications.* 1996, Rawat Publications: Jaipur.
7. Jones, T., *Corporate Killing. Bhopals Will Happen.* 1988, London: Free Association Books.
8. *Bhopal Gas Tragedy Relief and Rehabilitation Department* http://www.mp.nic.in/bgtrrdmp/setup.htm. 2004, Government of Madhya Pradesh.

9. Jaising, I., *Legal let-down*, in *Bhopal: The Inside Story*, T.R. Chouhan, Editor. 1994, The Apex Press: New York.

10. Jaskowski, J., et al., *Compensation for the Bhopal Disaster*. Int Perspectives in Public Health (Buffalo, NY: Ministry of Concern for Public Health), 1996. **1996**(11–12): p. 23–28.

11. Cassels, J., *The Uncertain Promise of Law: Lessons from Bhopal*. 1993, Toronto: University of Toronto Press Inc.

12. *Compensation Disbursement. Problems and Possiblities*. 1992, Bhopal Group for Information and Action: Bhopal.

13. *Commitments for Sustainable Development. Trade Unions at the Commission for Sustainable Development. Special "Business and Industry" Segment*. 1998, ICFTU: Brussels.

14. Morehouse, W., *The Ethics of Industrial Disaster in a Transnational World: The elusive quest for justice and accountability in Bhopal*.

15. Lapierre, D. and J. Moro, *It Was Five Past Midnight in Bhopal*. 1st Indian ed. 2001, New Delhi: Full Circle Publishing. 376.

16. *The Bhopal Gas Tragedy 1984– ? A report from the Sambhavna Trust, Bhopal, India*. 1998, Bhopal People's Health and Documentation Clinic: Bhopal.

17. Ramaseshan, R., *Profit Against Safety*, in *Bhopal: Industrial genocide?* 1985, Arena Press: Hong Kong.

18. Chauhan, P. S., *Bhopal Tragedy. Socio-legal Implications*. 1996, Rawat Publications: Jaipur.

19. Mathew, J., *Fighting for a cause*, in *Chronicle*. 1999, May 9: Bhopal.

20. Chadha, C.S., *Brief descripiton of work done for Bhopal gas tragedy relief. Personal communication*. 1995, Jan 9: Bhopal.

21. *The Truth of Bhopal. Bhopal Gas Tragedy, 15th anniversary*. 1999, Bhopal Gas Pidit Mahila Udyog Sanghthan, Stationery workers Union, Gas Pidit Avam Nirashrit Pension Bhogi Sangharsh Morcha: Bhopal.

22. Gunnarsson, B., *The fight for justice*. South Asia - Political and cultural magazine. Sweden, 1985(3): p. 3–6. [In Swedish]

23. Eckerman, I., *Long-term Health Effects of Exposure to the Bhopal Gases. Observations of the hazardous effects on the health of particularly vulnerable groups*. 1995, Nordic School of Public Health: Goteborg, Sweden.

24. *Gas victims brave dry taps in filthy colony*, in *Hindustan Times*. 2001, Jan 22: Bhopal.

25. Acquilla, S. and R. Dhara, *Interview with Dr. Banerjee and Moina Sharma , Center for Rehabilitation Studies*. 2004, Personal communication.

26. *Environment to the Ground and Bottom - experiences from the Halland ridge. Final Report from the Tunnel Commission*. 1998, The Ministry of Environment: Stockholm. SOU 1998:137 [In Swedish].

27. Srishti, *Surviving Bhopal. Toxic present - toxic future. A report on Human and Environmental Chemical Contamination around the Bhopal disaster site*. 2002, The Other Media: Delhi.

28. *Welcome to Bhopal and the Sambhavna Clinic*, in *Material for the Visit of Dominique Lapierre and Javier Moro*. 2001, Sambhavna: Bhopal.

29. *Foul Debris. The UCIL plant is still a health hazard*. Down To Earth, Dec 15, 2003.

30. *Presence of Toxic Ingredients In Soil/Water Samples Inside Plant Premises*. 1989, Union Carbide Corporation: USA.

31. *The story too horryfying to tell in a newpaper*. 2003, Bhopal International Coalition for Justice in Bhopal www.bhopal.net.

32. *The Bhopal Legacy. Toxic contaminants at the former Union Carbide factory site, Bhopal, India: 15 years after the Bhopal accident*. 1999, University of Exeter: Exeter.

33. *A Report on Mercury Contamination of Groundwater near the Union Carbide Factory at Bhopal*. 2001, Peoples' Science Institute: Dehra Doon.

34. *Hinterland. A Special Issue on The Bhopal Gas Tragedy*. 2003, Department of English, Hindu College: New Delhi. p. 34.

35. Morehouse, W. and A. Subramaniam, *The Bhopal Tragedy. What really happened and what it means for American workers and communities at risk*. 1986, New York: The Council on International and Public Affairs.

8.4 Medical Treatment and Rehabilitation

8.4.1 Health care infrastructure

8.4.1.1 Governmental health care

In the immediate aftermath of the Bhopal gas disaster, the health care system became tremendously over-loaded.

Within weeks, the State Government established a number of hospitals, clinics and mobile units in the gas-affected areas. Towards the end of 1986, 18 medical institutions were working in the gas-affected areas: 12 under the Chief Medical and Health Officer (Gas Relief), five under the Indian Red Cross Society, and one eye hospital under the Royal Commonwealth Society. The hospitals were upgraded with more beds and better equipment.

In 1984, none of the medical colleges in Madhya Pradesh had a psychiatrist on their faculty [1]. The administrators and the medical professionals considered that the complaints, especially the psychiatric symptoms, were imaginary and compensation related. From February 1985, teams of psychiatrists, clinical psychologists and social workers from Lucknow were located in Bhopal. Later, a training programme on recognising and handling mental health problems was offered to the medical officers.

No infrastructure for alternative abortion or ultrasonography was established [2].

For administrative purposes, the whole of the area affected by gas has been made into a separate medical district under the control of a Chief Medical Officer [3].

The Government of India has focused primarily on increasing the hospital-based services for gas victims. In addition to the already existing hospitals with a total of 275 beds, five additional hospitals with a total of 740 beds were built. The Pulmonary Medicine Centre was completed in 1994, but till 1998, had no permanent staff [4]. The Indira Gandhi Hospital for women and children was completed in 1994, but still did not function properly till 1998 [4]. In spite of this, it was inaugurated that year, probably to release the funds for the second action plan (see 8.3.1). The construction of the 540-bed Kamla Nehru Hospital has been going on since 1987 and the expenditure is nearly three times the budgeted amount [4]. It has been transferred from the Department of Gas Relief to the Department of Medical Education. In the interim, the survivors are being offered the 80-year-old 300-bed Sultania Zenana Hospital [5].

In 1994, there were approximately 1.25 hospital beds per 1,000 in Bhopal, which compares favourably to the recommendation made by the World Bank of 1.0 beds per 1,000 in developing countries [6]. Around 4,000 persons per day visited government hospitals.

In 2003, there were 65 specialists and 153 medical officers in various governmental hospitals, clinics and polyclinics, and in addition, 439 paramedical staff. The Chief Medical Officer, Gas Relief department, denied that there was a shortage of staff, or that ward boys were administering injections (*Hindustan Times,* February 4, 2003).

The hospitals received new equipment after the leak. However, there is a lack of people who know how to use it. Much of the equipment is dysfunctional, or accessories are missing [4, 5]. In the Pulmonary Medicine Centre alone there is equipment worth Rs 1.25 crore that is lying unutilised since the time it was purchased.

The primary health care system is not very well developed. A community health programme was started in the wards on an experimental basis in 1991–1992 [7]. Under it every individual in the ward got a health card so that the institution could regulate the treatment of a given group of persons. In 1994, the government ran two clinics and eight dispensaries. In the Action Plan submitted by the state government for the granting of funds by the centre, allocations for community health services made up only 2 percent of the total budget [4, 5]. The dispensaries were going to be handed over to the Bhopal Memorial Hospital Trust.

There is one tuberculosis hospital in Bhopal, in addition to the District Tuberculosis Centres. The National Tuberculosis Control Programme should be followed here. It includes free treatment for at least six months for everyone who is sputum positive. A follow-up by Sambhavna community workers showed that it was common for the patient having to buy at least one of the medicines, and many being denied medicines. As the hospital is open only a few hours every morning, it is very difficult for working people to reach it. It requires considerable effort for a gas victim to get treatment for TB [4].

Three projects under the Integrated Child Development Scheme (ICDS) were sanctioned in the affected areas to look after infants and children in the age group of six months to six years as well as nursing and expectant mothers. Under the auspices of these projects, 633 (792 according to the Government of MP [8]), Aanganwadi Centres were established for the care of infants, children, nursing mothers and expectant mothers. The beneficiaries received a daily supply of nutritious bread, medical and health check-ups and immunisation. On an average, 71,280 beneficiaries were provided milk free of cost daily [8]. These centres also provided non-formal pre-school education [9].

8.4.1.2 Bhopal Memorial Hospital Trust

Union Carbide was directed by the Supreme Court to finance a 500-bed hospital for the medical care of the survivors, including a 30-bed cardio-thoracic surgery unit and a research unit [4]. A large sum of money was set aside for the Bhopal Memorial Hospital Trust (BMHT).

The trust has now established a 350-bedded super speciality hospital (Bhopal Memorial Hospital and Research Centre, BMHRC) on land donated by the Government of MP, 8 km away from the gas-affected area. The hospital was inaugurated in September 1998. Because of the lack of personnel and equipment, there was still only rudimentary activity in 1999. The first mini-unit (outreach health centre) started in April 1998. The eighth and last started at the end of 2002.

In the Annual report 2001 [10], the hospital departments describe their activities in 190 pages. It is stressed several times that the aim of the hospital is to be the largest and best equipped hospital in Central India. The activities at the outreach clinics, that meet the major part of the gas affected patients, are described in 11 pages.

The hospital now has 14 departments, including psychiatry. Here is performed open heart surgery as well as hemodialysis of kidneys. Missing major specialities are gynaecology, obstet-

Bhopal Memorial Hospital and Research Centre-BMHRC (Christian Saltas)

rics and paediatrics. To save its resources for those patients who need it, it only sees patients who are referred from the outreach clinics. During the period July 2000 to December 2001, it had admitted 3,693 patients and carried out 2,202 surgical operations. A new research wing has been added, and there are plans to tie up with ICMR.

Each centre is manned by 4–5 medical doctors including an ophthalmologist. There are facilities for sophisticated ophthalmic examinations, an auto-analyser for blood examination and an X-ray machine. About 100–200 patients per day visit each clinic. In 2001, 389,000 gas tragedy victims were treated at these centres.

In the report, the different departments describe their academic activities, like bedside teaching, case presentations and seminars. Nothing of the sort seems to exist for the outreach clinics. Plans for the epidemiology and research wing had not materialized in 2003.

From the start, each patient was registered and for the use of the clinic, registration data was computerised. Each patient got a smart card for identification. Patient records were not used. All data on symptoms, examinations and treatments were written in the health booklet that was given to every patient. Later, it is said that the health books were taken back by the clinics. Now all clinical data are computerised and stored at a central server that can be reached from each outreach clinic.

Interviews with the doctors at the clinics showed that they had no special knowledge about the health problems caused by the gases. They treated the gas-affected patients the same way as they treated other patients. The resources at the clinics were sparsely used for the visiting patients. Medicines were provided free, except uncommon medicines that were not stored in the clinics. In one of the clinics, rational prescription was the guideline. No community health work, e.g. health education, was performed.

In a study by Sambhavna in 1998 [3], it was found that the physical examination was minimal or absent. Few of the patients were given any laboratory tests. In another study [11], it was found that drugs were prescribed for short-term symptomatic relief of non-specific symptoms. There were no treatment protocols for specific gas-related illnesses, and there were no mechanisms for following-up on their patients.

In 2000, a study was made where prescriptions issued in one BMHT clinic were compared with the description of the patients' symptoms in their health books [3, 12]. Only in 17 percent of the prescriptions, had the drugs given been properly selected and was the dosage appropriate.

The care at the hospital is supposed to be free for gas victims. In 2002, it was clear that at least one gas victim had paid for his treatment. The BMHT hospital would have to pay the money back.

The cardiology department of BMTHRC points out the probable paucity of funds in the future, with the declining interest rates [10]. They have undertaken publicity drives in the city and surrounding areas in order to rope in pay patients. Although the number of pay patients treated is much lower than the number of treated gas victims, the income from the pay patients exceeds the budget allocated for the gas victims.

The money available for the BMHT hospital and clinics is 86 million US$. The BMHT clinics and hospital are only obliged to provide free care for eight years. Maya Shaw [3] suggests that the remaining money should be used in ways that will have a more lasting impact on the communities, for example, health education and improvement of the community infrastructure to reduce the spread of infection.

8.4.1.3 Private doctors

Since the leakage, a very large number of private practitioners have opened clinics in Bhopal [13]. Of the private doctors in the severely affected areas, nearly 70 percent do not appear to be professionally qualified [4]. Most doctors of governmental hospitals practise privately [14].

8.4.1.4 Non-governmental organisations

Radical health groups, such as the Drug Action Forum and the Medico-Friend Circle, provided medical staff for the JSK (the People's Health Centre) set up by the combined opposition in June 1985. It was a free outpatient clinic working with education and public health. In June 1985, the clinic was raided and many of the staff were arrested. The data collected on the thiosulphate treatment (see 7.4.5) was confiscated and has still not been returned. Because of overloading and trouble with the authorities, the clinic was closed down after a few years.

Mother Theresa's Missionaries of Charity were reported to be operating a treatment camp at the Bhopal railway station for UCIL [2].

In 1985, the Indian Red Cross Society was given 5 million dollars by UCC as part of the interim relief. Four community clinics were set up in the gas affected area [4]. Four years later, UCC had the Indian Red Cross hand over all its remaining funds, and the clinics were closed down, the last one, around 1995. They are now prepared for transfer to the Bhopal Hospital Trust.

8.4.1.5 Sambhavna Trust

The Sambhavna* Trust was registered in June 1995 as a charitable trust with objectives concerning the welfare of the survivors of the Bhopal gas disaster through medical care, research,

---

* "Sambhavna" is a Sanskrit/Hindi word which means "possibility". Read as "sama" and "bhavna" it means "similar feelings" or "compassion".

health education and information dissemination [4, 15]. The trust collects its funds mainly through small donations from a large number of individuals. Appeals for funds are publicised through newspapers in India and abroad. Another source of income is the royalty from Lapierre's book [16].

Through The Bhopal People's Health and Documentation Clinic situated close to the affected communities, the Trust provides medical care, health education and research and monitoring facilities to the survivors free of charge. The clinic opened in 1996.

The clinic is directed by six national trustees – doctors, scientists, writers and social workers. Decisions are made through a consensus among the members of the board. An International Advisory Group provides long-term support. A 14-member panel of medical advisors from different countries provides technical support as needed. International and national volunteers give support, and it is co-operating with survivors' organisations as well as national and international non-governmental organisations.

The staff has grown, and was around 25 in 2003. There are three doctors of modern (allopathic) medicine, one of whom is a gynaecologist and one a psychiatrist (two hours a week), one doctor of Ayurvedic medicine, one Yoga teacher, two massage therapists and three community workers. Basic laboratory facilities for the examination of blood, urine, sputum and smears are available. The dispensary provides modern as well as Ayurvedic medicines free of charge. Rational medicines are used, and those with potential serious side effects (corticosteroids for systemic use, Depo-Provera) are not prescribed. A community research team collaborates with university institutions. The work is thoroughly described in the 1998 report [4]. Between 50 and 150 patients come here every day. Yoga and health camps have been held, as well as seminars for doctors.

Patients with diabetes, hypertension and other chronic diseases come every 15th day to see the doctor and to get their medicines. Some patients with severe respiratory problems are coming for bronchodilator and/or oxygen inhalation. Patients with chronic bronchitis may use Ayurvedic medicines during periods when they have less symptoms. Patients with skin problems are referred for Ayurvedic treatment.

Patients waiting outside Sambhavna (Christian Saltas)

Identification, diagnosis, treatment, follow-up and health education with regard to tuberculosis (both in the communities and in the clinic) is done by the health workers in collaboration with the WHO sponsored DOTS centre. Community health workers also carry out health education on reproductive health in five severely affected communities. Other community work involves creating awareness about the polluted ground water.

The overall impression of the staff is that a few patients go to other doctors as well as to Sambhavna. As they do not feel well treated at the hospitals, it is sometimes difficult to motivate them to go there when referral is necessary.

An American doctor, who volunteered to work at Sambhavna, has pointed out that working under these conditions, with small resources, 5–10 minutes per patient, who often has multi-symptomatology, is not easy [17, 18]. He observed that although examples of poor allopathic practice are common in Bhopal (as in almost any other city in the world), it is not clear that allopathic care at the clinic is better than the average for Bhopal. Although unique gas-related morbidity, particularly respiratory, psychiatric and gynaecological pathology, remains strong in Bhopal, the majority of diagnoses in this community are common ones and not necessarily gas-related.

The demands for the services are growing, and expansion is desirable. A new building, closer to the gas affected area, is going to be built in 2004. A garden for (ayurvedic) medical herbs has been laid out.

A small clinic has opened in the water polluted area close to the Union Carbide area. In Oriya basti, a community centre with a school has opened with support from Sambhavna.

8.4.2 Treatment

The ICMR working manuals of 1986 [19] and 1989 [20] gave advice on management of treatment.

For respiratory symptoms, bronchodilators were recommended for those patients who showed evidence of reversible airway obstruction. Corticosteroids might be used "as bronchodilators" for long-term treatment if a significant benefit was likely. Long-term side effects are described. Antibiotics were recommended for 6–8 day courses as soon as the sputum became purulent, and as prophylactics during acute viral chest infections in-patients with badly damaged lungs. Other drugs recommended were anti-anaemic drugs, analgesics, vitamins, cough suppressants and expectorants.

Recommendations for children include bronchodilators, steroids for acute episodes and for long-term treatment, sodium chromoglycate, antibiotics, expectorants, mucolytics and high doses of vitamin C.

A tuberculosis treatment programme was included in the manuals. It gave different alternatives for treatment lasting at least six months.

For psychiatric problems, anxiolytic drugs and antidepressants were recommended for both children and adults. There was no warning about the side-effects of long-term use of diazepam.

Non-pharmaceutical treatments were also recommended in the manuals:
• Respiratory exercises, physiotherapy and postural drainage for respiratory problems;

- Psychotherapy for psychiatric problems;
- Balanced diet;
- Physical exercises and yoga;
- Health education ("The patient must stop smoking for life");
- Social, economic and occupational rehabilitation.

However, there do not seem to be any resources provided for either physiotherapy or yoga, nor for psychotherapy and health education. The available social, economic and occupational rehabilitation is described in chapter 8.3.

In 1990, a study on 522 patients at two government hospitals for gas victims found that patients were being prescribed irrational or unnecessary medicines as well as medicines known to be hazardous, having been banned in other countries [4].

In the study by IMCB in 1994 [21], it was found that the therapies prescribed seemed to be aimed at giving temporary symptomatic relief rather than long-term amelioration of a chronic disease process. The patients reported the best effect on respiratory symptoms from bronchodilators and corticosteroids, but only a minority of patients with symptoms in the lungs received bronchodilators. In the government hospital sample, complaints of breathlessness, cough, chest pains and colds were all treated primarily with antibiotics, analgesics and anti-histamines. Vitamins were routinely prescribed as well as tonics and enzymes.

Drugs with a significant potential toxicity were prescribed in preference to less toxic alternatives. Analgin and hydroxyquinolin have been banned in many countries. Oral corticosteroids were used for rashes and as primary therapy for pulmonary symptoms. Often fixed dosage combination medicines were given; some in combinations that have the potential to induce serious toxicity. Several drugs prescribed for the gas victims have the potential to induce significant iatrogenic disease [21].

Drugs used commonly are analgesics, bronchodilators, steroids, antacids and H_2 receptor antagonists (Ranitidine). Chloromycetin ointments and sulphacetamide eyedrops have been liberally used [4]. The private doctors prescribe steroids, intravenous drips and antibiotic injections indiscriminately [4].

In the 1996 Sambhavna study of 50 chemist shops, it was found that the quality of the prescriptions was low [4]. The three most sold drugs were anti-infectives, enzymes/tonics and cough syrups. The following groups had an "unknown composition": Ayurvedic drugs, analgesics, steroids and anti-ulcer drugs. Useless, hazardous and combinations of irrational drugs were mostly given for ill-defined symptoms, and appropriate drugs were used more for well-defined symptoms or conditions.

In the 1998 study of prescriptions from the BMHT mini units [11], cough was often treated with anti-histamines and anti-infectives. Psychiatric symptoms were commonly treated with vitamins and benzodiazepemes. No drugs specific for depression of PTSD were found.

Drugs are supposed to be provided free of charge at the government hospitals, the TB-hospital and the BMHT clinics. However, in 1998, 30–40 percent of the medicines were out of stock in two hospitals [4]. This verifies what the patients say, that they have to buy medicines outside the hospitals [22].

There is no information on the efficacy of the treatments at government and private hospitals or clinics, as no evaluation has been made. At Sambhavna, it has been documented that a substantial number of persons have not found relief through the treatments, and many report a relapse of symptoms [4]. This is true also for patients receiving Ayurvedic treatment.

In 1997, at Jawaharlal Nehru Hospital close to the most affected area, 80 percent of the medicines meant for delivery free of charge were out of stock [5].

At Sambhavna, patients with severe lung problems are treated with bronchodilators and local corticosteroids in aerosols. Some are coming for regular treatment with nebulisator, and a few also gets treatment with oxygen. It is not known to what extent this is done in other hospitals and clinics. Patients with chronic bronchitis may use Ayurvedic medicines during periods with less symptoms. Yoga has documented effects for many patients [23]. Patients with skin problems or obstipation are referred for Ayurvedic treatment, as well as young girls with menstrual problems. Patients with diffuse symptoms like pains "everywhere" and anxiety are often referred for complementary medicine (Ayurvedic treatment, massage, yoga) and are content with this. Some of these patients are referred to the psychiatrist, who prescribes anxiolytic and antidepressive medicines.

In 2002, facilities for screening, diagnosis and treatment of cervical cancers were unavailable at the hospitals designated for survivors [24]. In Sambhavna, cytology and colposcopy was introduced during 2003.

Dr Banerjee at the Centre for Rehabilitation Studies stated in 2003 that the ICMR manual for treatment of gas victims does exist but is not being followed in the governmental hospitals [13]. There is indiscriminate prescription of steroids by private practitioners for the lung condition which may be, in part, responsible for the gastrointestinal problems.

8.4.3 Documentation

During the first days after the leakage, it was impossible for the medical staff to organise any kind of documentation. Nor was it done later. In December 1985, it was reported that proper medical records only existed for less than 10,000 victims [2].

In 1999, these conditions had not changed much. In the government hospitals, the patients gathered in the emergency departments, and were treated only if they had acute complaints. If a record was written, it was placed on the top of the day's records and later put in some storeroom. In the government hospitals and clinics, it is not possible to follow the development of the health of an individual patient. Sometimes the patient is given the records.

Private doctors often give the records and results to the patient.

In Sambhavna, everything is documented in two ways: in a health booklet for the patient, and visit cards for the clinic. The latter are all collected in an individual envelope, with a unique code for each patient. Administrative as well as medical data is entered in a computer program (EpiInfo). It is analysed in EpiInfo, Excel and FoxPro.

In the BMHT clinics, all medical documentation was originally entered in a booklet that the patient carried him/herself. Only administrative data was entered in a computer program. Later this was changed. The health booklets were withdrawn, and all clinical data were computerised.

8.4.4 Survivors' views

Many survivors express a common dissatisfaction with the health care facilities [6, 22, 25, 26]. They pay a lot of money, but they are not cured. Often they have to pay for medicines that should be dispensed free. Other reasons for dissatisfaction with hospital services are overcrowding and long waiting times, lack of proper attention, distance and the unprofessional behaviour of doctor/staff. The complaints are similar to those expressed by other slum dwellers [27].

Those who can afford it, prefer private doctors to the government hospitals, as they consider that they are given a more patient hearing there [22, 28]. In surveys it was found that 77 percent of the persons undergoing medical treatment visit private clinics [4].

Survivors' organisations have long been demanding the setting up of a community based health infrastructure in which medical care and monitoring services are available to persons within the community itself and the hospitals are used as referral centres [4].

In letters, the survivors' organisations have opposed the building of the Bhopal Memorial Trust Hospital, and said that "in the absence of any information based treatment protocol being offered, the Trust's hospital will be just one more building, good to look at, but ineffective in care" [4]. They are calling for the setting up of a competent local body with survivors' participation that would take control of the funds and administer appropriate health care among the gas affected population.

8.4.5 Recommendations from IMCB

The International Medical Commission on Bhopal (IMCB) has made recommendations on medical, social and environmental rehabilitation. It has proposed a four-tier health infrastructure with community health units as the base in place of the current hospital centred system [6].

IMCB has also expressed serious reservations regarding the relevance of the special facilities in the Bhopal Memorial Trust Hospital.

IMCB has recommended that the Government establish an independent National Medical Commission in Bhopal, and convene an International Conference on Urban Community Health Centres.

8.4.6 Comments

Health care in Bhopal has many shortcomings that would not be too difficult to improve:

- There is no formal infrastructure at the community level, and there are no concrete plans for improving the community based health care system;
- The working manuals of ICMR were poorly distributed to the doctors;
- Neither government nor private doctors use a rational prescription of drugs;
- The stores of rational drugs at hospitals and clinics are not adequate;
- A systematic further education programme for graduate and non-graduate doctors is missing;
- Alternative and non-pharmacological treatments are not evaluated properly;
- There is no systematic documentation of symptoms, investigations and treatments;
- The inhabitants in the *bastis* are provided with hardly any systematic health education.

In rich countries, patient records usually belong to the hospital or clinic, and there are laws regulating responsibilities and storage. In Bhopal, it seems safer to give the documentation to the patient than to keep it at the hospital. But this will inhibit clinical research. To get the final compensation, the victims have had to deliver their medical records. If copies were not made, the patient would have lost the information about him/herself.

Survivors and activists are very sceptical of the BMTHRC hospital, as they believe it will be used mainly by Bhopal's upper class. The future will show whether they are right.

Among the victims and activists, there has been a wide-spread idea that if UCC released information about the content of the gas cloud, it would be possible to come up with an effective course of treatment. But when human tissues are seriously wounded, they heal with scars. This is the case also with the internal tissues. With the best and immediate treatment, the internal injuries in lungs, eyes, nervous system etc., would only have been mitigated, not cured. The only treatment that might have had good effect on certain injuries would have been sodium thiosulphate (see 7.4.5). Today, only symptomatic treatment is possible, as for other chronic diseases.

If health education had worked all these years, the symptoms might have been somewhat mitigated. But it is still not too late for the smokers to stop smoking, for non-smokers not to start, for housewives to clean the drinking water, or for infected persons to reduce the spreading of the infection. If only someone would tell them ...

Better access to health care also increases the demand. People will seek help from professionals for problems they earlier thought were inescapable or natural. When health care is free, more people will use it also for minor problems that they earlier treated themselves. That there is an increasing number of patients at hospitals and clinics cannot be used as a proof that the health problems in the population are increasing. Demand is not always the same as need.

However, we know that gas victims have uncommon chronic diseases. It is reasonable to take it for granted that there is an over morbidity also in common diseases (see 8.1.5). If also those who were infants at exposure should get adequate health care for as long as they live, the health care facilities for gas victims should be developed and maintained at least for an additional two generations. This would also be an opportunity to follow up on the children of gas victims.

8.4.7 References

1. Murthy, R.S., *Mental Health Impact of Bhopal Gas Disaster*. 2002, The Other Media: New Delhi.

2. Jones, T., *Corporate Killing. Bhopals Will Happen.* 1988, London: Free Association Books.

3. *Welcome to Bhopal and the Sambhavna Clinic,* in *Material for the Visit of Dominique Lapierre and Javier Moro.* 2001, Sambhavna: Bhopal.

4. *The Bhopal Gas Tragedy 1984– ? A report from the Sambhavna Trust, Bhopal, India.* 1998, Bhopal People's Health and Documentation Clinic: Bhopal.

5. *13th Anniversary Fact Sheet on the Union Carbide Disaster in Bhopal,* in *Compensation Disbursement.* 1997, Bhopal Group for Information and Action: Bhopal.

6. Verweij, M., S.C. Mohapatra, and R. Bhatia, *Health Infrastructure for the Bhopal Gas Victims.* Int Perspectives in Public Health (Buffalo, NY: Ministry of concern for publich Health), 1996. **1996**(11–12): p. 8–13.

7. *Socio-economic Impact of Disbursement of Interim Relief to Gas-affected Families of Bhopal.* 1991, Academy of Administration, Government of Madhya Pradesh: Bhopal. p. 154.

8. *Bhopal Gas Tragedy Relief and Rehabilitation Department* http://www.mp.nic.in/bgtrrdmp/setup.htm. 2004, Government of Madhya Pradesh.

9. Chauhan, P.S., *Bhopal Tragedy. Socio-legal Implications.* 1996, Rawat Publications: Jaipur.

10. *The Bhopal Memorial Hospital and Research Centre and Outreach Health Centres. Annual Report 2001.* 2001, Bhopal Memorial Hospital Trust: Bhopal.

11. *Assessment of treatment offered at the Bhopal Hospital Trust's community clinic no.1 and analysis of Bhopal Hospital Trust prescription data.* 1998, Sambhavna Clinic: Bhopal.

12. *Union Carbide Disaster in Bhopal. Fact Sheet. 16th Anniversay of the December 2–3, 1984.* 2000, Bhopal Gas Pidit Mahila Udyog Sangathan and Bhopal Group for Information and Action: Bhopal.

13. Acquilla, S. and R. Dhara, *Interview with Dr. Banerjee and Moina Sharma*, Center for Rehabilitation Studies. 2004, Personal communication.

14. *Serving the symptom. The government still does not know what still afflicts people in Bhopal.* Down To Earth, 2003(December 15): p. 29–32.

15. *Stop the Ongoing Medical Disaster. Bhopal healthcare is sick.* 2000, Sambhavna Trust: Bhopal.

16. Lapierre, D. and J. Moro, *It Was Five Past Midnight in Bhopal.* 1st Indian ed. 2001, New Delhi: Full Circle Publishing. 376.

17. Carlsten, C., *Critical Issues in Diagnosis and Treatment of Common Presentations in Bhopal India.* 2000, California Academy of Family Practice and Sambhavna: Bhopal.

18. Carlsten, C., *An Eyewitness in Bhopal, 15 years after the disaster.* BMJ, not published, 2000.

19. *Working Manual 1 on the Health Problems of Bhopal Gas Victims. Assessment & management.* 1986, April, Indian Council of Medical Research: New Delhi.

20. *The Health Problems of Bhopal Gas Victims. Assessment & Management. Working Manual 2.* 1989, Dec, Indian Council of Medical Research, DST Centre for Visceral Mechanism.: New Delhi.

21. Bhatia, R. and G. Tognoni, *Pharmaceutical use in the victims of the carbide gas exposure.* International Perspectives in Public Health (Buffalo, NY: Ministry of Concern for Public Health), 1996. **1996**(11–12): p. 14–22.

22. Eckerman, I., *The health situation of women and children in Bhopal.* International Perspectives in Public Health (Buffalo, NY: Ministry of Concern for Public Health), 1996. **1996**(11–12): p. 29–36.

23. Gupta, A., S. Durgavanshi, and I. Eckerman. *Effects of Yoga practices for respiratory disorders related to the Union Carbide Gas Disaster in 1984.* in *The XVI World Congress of Asthma.* 1999. Buenos Aires.

24. Sarangi, S., *The Bhopal aftermath: generations of women affected,* in *Silent Invaders. Pesticides, Livelihoods and Women's health,* M. Jacobs and B. Dinham, Editors. 2003, Zed Books Ltd: London.

25. Narayan, T., *Submission on the health status and health care of victims of the Bhopal gas disaster of 1984, to the International Medical Commission on Bhopal.* 1994, Medico Friend Circle and Bhopal Group for Information and Action: Bhopal.

26. *Statement of Bhopal Peedit Mahila Stationery Karmachari Sangh before the IMCB.* 1994, Bhopal Group for Information and Action: Bhopal.

27. *Report on family welfare/primary health care needs of slumdwellers of Bhopal town and formulation of proposal for strengthening of existing family welfare services and creation of new health facilities.* 1992, Department of Preventive and Social Medicine, Gandhi Medical College: Bhopal.

28. Eckerman, I., *Long-term Health Effects of Exposure to the Bhopal Gases. Observations of the hazardous effects on the health of particularly vulnerable groups.* 1995, Nordic School of Public Health: Goteborg, Sweden.

8.5 Psycho-socio-economic Effects on Survivors

8.5.1 Who was affected?

One study found that 74 percent of those in affected areas fled on foot, 6 percent by vehicle (motorised or bicycle) and 21 percent remained. None of those who went by a vehicle died. A study done by the Centre for Social Medicine, Jawaharlal Nehru University, New Delhi, found that "those who died are the poorest. More than half the affected people belong to an income group – about 150 rupees per head per month – which cannot afford two full meals a day around the year. Those who died were even more disadvantaged than the overall affected population."

Henry Falk of the US Centre for Disease Control noted that people who were lucky enough to live in well-sealed homes managed to avoid the choking gas while some escaped by climbing to a higher level [1].

In 1981, 38 percent of the population were in the age group 0–14 years [2]. Small children were more severely hit than the adults were (see 7.4.1, 7.4.3, 8.1.2). If the total immediate death toll was 2,500, the figure for children would be at least 1,000. If we instead estimate the death toll to be 8,000, this would imply that 3,000 children died within the first weeks. The number of severely affected children would be 13,000, moderately affected 29,000, and mildly affected children, 166,000.

Some sources maintain that nearly half of those who died immediately were children [1]. That would mean that 2,000 to 4,000 children died directly.

8.5.2 Long-term economic effects

The economic situation of the most affected part of the population can be categorised as follows:
• Dependence on physical and casual work for income;
• Dependence on cattle for income;
• Dependence on sons for support in old age.

A door-to-door survey in the last week of December 1984 found that 75 percent of the workforce was incapable of work, mainly because of breathlessness [3]. Many families lost their cattle. Many parents and widows lost their sons and will thus have no support at all in old age.

An analysis of the socio-economic survey conducted between January 1985 and December 1985 revealed that about 25,000 families reported a total or substantial loss of income [4]. With a mean of five members per family, around 125,000 persons suffered from a reduced family income.

Young girls had difficulty in getting married, because of doubts whether they would bear healthy children [5]. When the interim relief began to be paid out, they again became interesting objects for marriage [6].

8.5.3 Long-term effects on family life

The survivors not only lost their economic standard; their whole existence changed. A witness said to me: "Death would have been a relief. It is worse to be a survivor. Our handicaps, like memory loss, are not visible to others." A man who lost both his sons said: "Who will care for us when we grow old? Our life has been destroyed." These sentences are repeated in many ways by hundreds of witnesses. The testimonies of survivors are well documented [7–15].

An unknown number of children became orphans. An orphanage was put up and around 20 children lived there for some time. Most of them later found relatives or others to care for them and left [16].

The loss of income not only meant a lower standard of living, but also involved worries about the future. Parents did not have sons to support them in old age any longer. Young girls would not get married and would remain dependent on their parents.

The loss of health not only led to restrictions in activities compared to previously, but also involved spending time at hospitals and clinics. The expenses for medicines could sometimes exceed expenses on food [7]. Adults worried about their own health as well as their children's, and pregnancies were avoided for many years [6, 7].

The conditions in the family changed. In many cases, some or even most family members were not there any more. Others might be ill, having not only somatic symptoms but also suffering from irritability, depression and mourning. Daughters, old enough to be married, were still around. Relatives who had lost their families moved in. There is a mean of one ill person per family [11], but with a tendency for concentration in certain families [6].

The social life outside the family changed; relatives, neighbours, or fellow workers were lost.

Fear led to preparedness to leave the town immediately when threats of new leaks were experienced (see 7.3.7, 8.6.1).

The victimisation and disempowerment of the survivors is discussed in sections 8.3.3.1, 8.8.4, 8.8.7 and 8.8.10.4.

8.5.4 Effects of interim relief

In a survey of 21,000 survivors, performed in the pre-disbursement period of interim relief in June and July, 1990, and repeated in December 1990 and January 1991, some short-term effects of interim relief were shown [17]. Effects regarded as favourable were the following:

1. *Expenditure of the families on education increased.* There was an improvement in the number of children going to school. More families could send their children to private schools and spend more money on better education for their children.
2. *People could spend more money on their treatment.* They have also indicated a shift towards more reliance on private medical help as against the use of government facilities. This is indicative of their capacity and willingness to seek individual and better attention, even though it may be rather costly.
3. *Expenditure on food also significantly increased.* The interim relief enabled the beneficiaries to spend more on food and nutritive food.
4. *Their housing conditions also improved,* both in terms of accommodation and the available nature and degree of amenities.

Other effects were regarded as unfavourable:

1. *Undesirable expenditure.* Sudden availability of extra funds with the people, especially in the hands of the comparatively poorer sections of the society, gave rise to expenditure more on non-essential and wasteful items rather than on necessary items. It has been noted that family expenditure on liquor, gambling and various luxuries increased significantly.

2. *Intra-family relations.* While family relations with their neighbours and the community improved, intra-family relations became strained. This was presumably the outcome of the additional money available. Earning members tended to contribute less to the family and spend more on their personal pleasures.

3. *Status of women decreased.* An interesting aspect of social impact was noticed in the case of status of women. Interim relief tended to adversely affect the status of women in as much as the number of female heads of families declined in the post-disbursement period. It seems that in these families, women were earlier earning members and were heading the family; subsequently, interim relief provided income in the hands of male members who established their dominance and became heads.

8.5.5 Comments

How seriously a person was affected was largely dependent on socio-economic factors:
- Age;
- Living area;
- Type of house;
- Access to a vehicle.

The psycho-socio-economic long-term effects might have been mitigated through investments of a different kind:
- Involving the survivors in the decision-making process;
- A community based health care system of good quality;
- Economic support immediately after the leak;
- Permanent work-sheds for those who cannot find suitable work on the open market;

- Pension and boarding for those who cannot provide for themselves and have no family to provide support.
- A programme for systematic debriefing work, including psychological support;
- An environmental rehabilitation programme for the most affected areas.

When Mr Ram Shankar Tiwari, professor of economics at the Academy of Administration and the person responsible for the survey on interim relief, was interviewed in *The Chronicle*, 28/11/1994, he stated: "There was no definite improvement in the peoples' standard of living." He also stressed the findings that money was "frittered away on things like the consumption of liquor and gambling". He said that the money has certainly not helped to bring about a better quality of life for the citizens. Now, when the final compensation is about to be paid out, "Bhopal's economy is set to suffer total disruption and chaos. This huge amount of money will make life miserable. Except for human beings, everything else will become costly here."

In my opinion, the effects of the interim relief seem natural. If poor people get more money, they do as the rich do: spend them on good and bad things. Sending more children to school, spending more money on housing, food and medicines and having the possibility of buying non-essential items, must be regarded as a better living standard. To be able to work less is a positive thing, specially for teenagers and old people (see 8.6.3). To get a higher salary when you are working must be a positive factor, from the individual's point of view.

The study shows how important it is for the situation of women to have an income of their own.

8.5.6 References

1. Jones, T., *Corporate Killing. Bhopals Will Happen*. 1988, London: Free Association Books.
2. *Census 1981*. 1982, Government of Madhya Pradesh, The Census Department: Bhopal.
3. *The Trade Union Report on Bhopal*. 1985, ICFTU-ICEF: Geneva, Switzerland.

4. Chauhan, P. S., *Bhopal Tragedy. Socio-legal Implications*. 1996, Rawat Publications: Jaipur.
5. Sathyamala, C., *Fertility and gynaecological disorders: Impact of Bhopal gas leak disaster*, in *Dep of Epidemiology and Population Sciences*. 1993, London School of Hygiene and Tropical Medicine: London.
6. Eckerman, I., *Long-term Health Effects of Exposure to the Bhopal Gases. Observations of the hazardous effects on the health of particularly vulnerable groups*. 1995, Nordic School of Public Health: Goteborg, Sweden.
7. Eckerman, I., *The health situation of women and children in Bhopal*. International Perspectives in Public Health (Buffalo, NY: Ministry of Concern for Public Health), 1996. **1996**(11–12): p. 29–36.
8. *Bhopal Lives! Anniversary Notes*. 1994, Bhopal Gas Peedit Sangharsh Sahayog Samiti and Bhopal Group for Information and Action: Bhopal.
9. *Statement of Bhopal Peedit Mahila Stationery Karmachari Sangh before the IMCB*. 1994, Bhopal Group for Information and Action: Bhopal.
10. Narayan, T., *Submission on the health status and health care of victims of the Bhopal gas disaster of 1984, to the International Medical Commission on Bhopal*. 1994, Medico Friend Circle and Bhopal Group for Information and Action: Bhopal.
11. *The Bhopal Gas Tragedy 1984– ? A report from the Sambhavna Trust, Bhopal, India*. 1998, Bhopal People's Health and Documentation Clinic: Bhopal.
12. *Bhopal: The Second Tragedy*. 1995, Central Broadcasting: Birmingham.
13. *Voices from Bhopal*. 1990, Bhopal Group for Information and Action: Bhopal.
14. *Nothing to Lose But Our Lives. Empowerment to oppose industrial hazards in a transnational world*, ed. D. Dembo, et al. 1987, New York: Council on International and Public Affairs.
15. Mehta, S., *Bhopal Lives*. Village Voice, 1996(Dec 3).
16. Weir, D., *The Bhopal Syndrome. Pesticides, Environment and Health*. 1988, London: Earthscan Publications Limited.
17. *Socio-economic Impact of Disbursement of Interim Relief to Gas-affected Families of Bhopal*. 1991, Academy of Administration, Government of Madhya Pradesh: Bhopal. p. 154.

8.6 Effects on Society

8.6.1 Effects on the "spirit" of the town

It is natural that the leakage occupied the minds of people for some time afterwards. Bidwai writes one week after the event, "there is a total collapse of all individual identities into a general identity – victim of Union Carbide, direct or indirect, a participant in the collective trauma, the sufferer in an unending purgatory" [1]. However, the "victimi-sation" seems to have become permanent [2, 3].

The change was also physically noticeable. Before the leakage, all the boys would be out on the streets playing cricket. After the leakage, the streets were silent. No boys played cricket – they did not have the breath.

The day after the leakage, several thousand Bhopal residents tried to storm the factory. Plant officials and police guarding the plant only succeeded in turning the crowd away by telling them that another poisonous gas leak was in progress [4–6]. This led to people leaving Bhopal for neighbouring districts. "Operation Faith" (see 7.3.7) was the second exodus from Bhopal. A third exodus took place in March 1985.

The migration out of Bhopal was sparked by the fear of another leak, anxiety over soil and water contamination, and difficulties in earning one's living. Many of the migrants may have stayed away for a long time. There was also a return migration of people originally coming from other places. They left Bhopal to go "home" where they felt safer. There are no official figures of how many migrated, but it may be tens of thousands.

Immediately after the leakage, people of all kinds came rushing to Bhopal to help. After a few weeks, it was quite apparent that the government was not going to do anything that it was expected to do, both in terms of providing relief and rehabilitation and fighting for justice against an offending multinational corporation [7]. Soon the people from the *bastis* joined the outsiders, and local committees without elected leaders were formed. The year following the disaster, there were four active organisations in Bhopal. Although there were sharp differences among them, in May 1985, they decided to join together and run a clinic (see 8.4.1.4).

The residents, UCC workers and activists in Bhopal, protested at an early stage against the behaviour of the government. In 1985, state repression towards the opposition was very strong, which resulted in mass arrests and police violence (see 8.8.2). Since then, rallies and protest meetings have become a regular part of life. Every Saturday, the BGPMUS *(Bhopal Gas Peedit Mahila Udyog Sangathan)* arranges a meeting. Every anniversary, large meetings and rallies are organised. Foreigners are often seen taking part.

Economic compensation became a very important issue – even a small amount of money might make the difference between starving and satisfying one's hunger. It is said that a number of Bhopalis developed imaginary illnesses or injured themselves in order to claim compensation [5]. When the interim compensation was paid out, the girls became attractive as marriage partners [3].

According to Meenakshi [8], the Bhopal gas tragedy galvanised public interest groups. By 1989, most were either

Rally on December 3 (Christian Saltas)

Demonstration on December 3 (Christian Saltas)

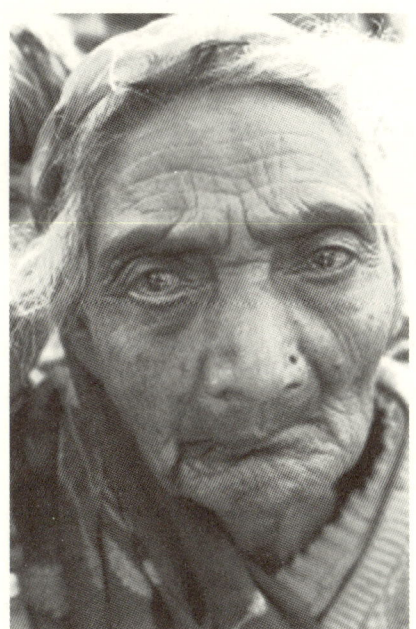

An old victim (Christian Saltas)

Today, it is the logotype of Dow that is painted on the walls (Christian Saltas)

defunctional or had narrowed their focus to highlighting the victims' plight.

Many misunderstandings and ideas have been circulated during the time since the disaster:

- That pockets of gas left behind by the cloud were continuing to poison people for years;
- That the water of the lake contained MIC;
- That an internal "cyanide pool" within the body continuously worsened the injuries;
- That if only information about the contents of the cloud was released, effective treatment to cure the injuries would be possible.

The male leader of the largest victims' organisation said that the organisation "has made the *burka*-clad Muslim women aware of their rights. They now know about their position in the society and are learning to be self-reliant" [9]. Another success is the absence of the feeling of untouchability among its members. There is a political awakening among them. During the political riots in 1992, many Hindu families of the organisation had helped their Muslim brothers by giving them shelter and food. In 2002, after the riots in Gujarat, there were no signs of animosity between Hindus and Muslims in Bhopal. Caste and religion are not big issues in Bhopal.

8.6.2 Effects on the infrastructure

The effects on the infrastructure and economic life are described by Chauhan [6], Cassels [10] and others.

Over half of the population of Bhopal was exposed to the gases. Those who were most affected were those who did the labouring work: road repairs, building, transport, industrial work, household work, small businesses. Many others migrated from Bhopal for economic reasons or because of fear. When all these people did not go to their jobs, the whole town was affected.

Communications came to a stop. Key-persons for train and public transports were missing. For example, 72 railway employees were killed and about 400 admitted to hospitals. For days, truck owners and drivers were afraid of making even a single trip to Bhopal.

The State Government ordered the closure of all schools, colleges, offices and shops until further notice. The closure of the fish and meat markets did not hit only the salesmen at the markets, but also butchers and fishermen outside Bhopal. The closure of the shops involved economic losses for the shop owners as well as for the contractors. Many shops never opened again.

At least 2,000 commercial animals were killed and crops were destroyed. With shops and markets closed, the prices of food went up.

When trade, commerce, building, and industrial activity came to a halt, it led to large-scale unemployment. It was not only the daily-wage labourers who were left without work. Around 850 workers had been employed at the UCIL plant in Bhopal. Only two of them died, but all lost their jobs. There was a complete collapse of the administrative machinery as well.

Many migrated from Bhopal, but others came in their stead: lawyers, volunteers, experts, media, doctors etc.

8.6.3 The effects of the distribution of relief

In January 1985, large protests were organised against the withdrawal of free distribution of rations [4]. This resulted in police violence and arrests. After this, the rations were restored.

A doctor from Bhopal describes the situation immediately after the leak [4]. Food, blankets, clothing, medicines were distributed to people who lived in the area, irrespective of whether they had suffered or not. Men stopped working, got a certificate from a local doctor and joined the handout queues. People outside the affected area resented the growing level of prosperity in the affected area. This brought about a split in the society. On several occasions, other residents stoned lorries carrying supplies to the affected areas.

In a survey of 21,000 survivors, performed in the predisbursement period of interim relief in June and July, 1990, and repeated in December 1990 and January 1991, some short term effects of interim relief was shown [11]. Favourable effects on society were the following.

1. *Social relations in the community*. There was an improvement in the relations of people with their neighbours. The number of quarrels in the locality were considerably reduced. There was a decline in various cognizable offences, particularly those relating to thefts and looting.
2. *Favourable macro effects*. Injection of additional purchasing power into the city's economy has increased the demand for various commodities and services. Job opportunities have improved and possibilities of self employment by using the relief amount as margin money have opened up. Further, although people were reluctant to give information about savings and investments, secondary sources of information do indicate that a part of the interim relief, though a very small part, has gone into savings in the form of recurring deposits in banks.

Other effects were regarded as unfavourable:

1. *Increase in prices*. Both wholesale and retail prices increased considerably. The increase in the prices of most essential

commodities in Bhopal during the period of the study was comparatively higher than in several other cities, including a few adjacent towns such as Sehore and Hoshangabad and the biggest city of M.P., Indore. It has been noted that the poorer sections of the people were also the worst affected by the gas tragedy. As a result, they were also hit the hardest by these price increases.

2. *Reduced incentive to work.* Availability of interim relief has adversely affected the incentive to work. The normal backwards sloping tendency of the supply curve of labour seems to have had its effect. Since the interim relief considerably increased family incomes (in the case of the lowest income group this increase was more than 100% of their income from regular sources), there was a tendency to put in less of an effort in work thereafter. This has been reflected in the reduced number of average working hours of the working members of the families as well as in the reduced overall employment. Shortage of workers has been particularly marked in domestic and industrial sectors. On the other hand, the number of daily casual workers has tended to increase. This indicates that some people have left their regular employment and are now only working on a daily basis as and when they feel the necessity to do so. Regarding the impact on unemployment, it was found that both among the teenagers and the aged, the number of unemployed persons increased after disbursement of interim relief.

3. *Higher wages.* As a result of the reduced incentive to work, the rates have tended to increase all around. Like prices, wages in Bhopal have also increased more steeply compared to several other cities during the same period.

4. *Certain macro effects.* An attempt was made to find out the effect of interim relief on interest rate structure and rent. However, no meaningful data was available on these aspects. Hence, these aspects could not be probed.

Other studies have indicated that living standards have increased since the gas leak, more in families where women had employment than in families where the women lost their jobs [3]. Children usually go to school and many are immunised. Much of the relief is used for medicines [2].

8.6.4 Effects on the political society

The Congress (I) Party got involved in the state repression of the opposition in Bhopal [4]. During the June 1985 repression (see 8.8.2), members of the party distributed posters alleging that the Morcha had CIA links. During the rally, they infiltrated the crowd and threw stones at the police.

Political decisions on relief were made to gain positive publicity or even votes (see 8.8.2).

The capitalist opposition parties did not really respond to the catastrophe and its aftermath [4].

Jones [4] notes the effects on the political society in Madhya Pradesh. The Minister for Labour in the MP government resigned after accepting moral responsibility for the disaster. Two labour ministry officials were suspended and two others sent on leave. The Chief Minister dismissed the chief inspector of factories, who "went on renewing the licence of the UC factory every year without taking into cognisance the safety lapses in the factory".

8.6.5 Comments

I have not found any attempt to calculate the total costs for the leakage. They must, however, be considerable, if the economic impact on society is included.

It is obvious that the struggle for justice in the form of survivors' organisations has contributed to the empowerment of women.

There have very little signs of animosity between Hindus and Muslims in Bhopal, unlike in Gujarat and some other parts of India. In the struggle for justice, they have worked side by side. They have a common history and a common enemy.

8.6.6 References

1. Bidwai, P., *The poisoned city - diary from Bhopal*, in *Bhopal: Industrial Genocide?* 1985, Arena Press: Hong Kong.

2. Eckerman, I., *The health situation of women and children in Bhopal*. International Perspectives in Public Health (Buffalo, NY: Ministry of Concern for Public Health), 1996. **1996**(11–12): p. 29–36.

3. Eckerman, I., *Long-term Health Effects of Exposure to the Bhopal Gases. Observations of the hazardous effects on the health of particularly vulnerable groups*. 1995, Nordic School of Public Health: Goteborg, Sweden.

4. Jones, T., *Corporate Killing. Bhopals Will Happen*. 1988, London: Free Association Books.

5. Lapierre, D. and J. Moro, *It Was Five Past Midnight in Bhopal*. 1st Indian ed. 2001, New Delhi: Full Circle Publishing. 376.

6. Chauhan, P., S., *Bhopal Tragedy. Socio-legal Implications*. 1996, Rawat Publications: Jaipur.

7. *Hinterland. A Special Issue on the Bhopal Gas Tragedy*. 2003, Department of English, Hindu College: New Delhi. p. 34.

8. Meenakshi, N., *Industrial disasters working towards oblivion*, in *India Disasters Report. Towards a Policy Initiative*, S. Parasuraman and P.V. Unnikrishnan, Editors. 2000, Oxford University Press: New Delhi.

9. Mathew, J., *Fighting for a cause*, in *Chronicle*. 1999, May 9: Bhopal.

10. Cassels, J., *The Uncertain Promise of Law: Lessons from Bhopal*. 1993, Toronto: University of Toronto Press Inc.

11. *Socio-economic Impact of Disbursement of Interim Relief to Gas-affected Families of Bhopal*. 1991, Academy of Administration, Government of Madhya Pradesh: Bhopal. p. 154.

8.7 Preventing "New Bhopals"

8.7.1 India

8.7.1.1 Laws

In 1972, after the UN Stockholm conference, the National Committee on Environmental Planning (NCEP) was set up. The Central Board for Prevention and Control of Water Pollution appeared in 1974. In 1980, the Department of Environment was formed, and in 1985, the Ministry for Environment and Forests (MoEF) was established.

After the Bhopal Gas Leak, many new acts and rules were created.

The following acts and rules lay down requirement for emergency preparedness and litigation [1–5]:

Acts:
* The Factories Act, 1948, as amended in 1976 and 1987;
* The Water (Control and Prevention of Pollution) Act, 1974;
* The Air (Prevention and Control of Pollution) Act, 1981;
* The Environment (Protection) Act, 1986;
* The Public Liability Insurance Act, 1991, as amended in 1992.

Rules:
* The Model Rules under the amended Factories Act;
* The Manufacture, Storage and Import of Hazardous Chemicals Rules (MSIHCR), 1989;
* The Public Liability Insurance Rules, 1991, as amended in 1992;
* The Emergency Planning, Preparedness and Response to Chemical Accident Rules (EPPRCA), 1999.

The Public Liability Insurance Act, 1991, was enacted to provide prompt relief to non-worker victims [6]. The delinquent units take insurance cover that provide prompt relief and compensation to third parties for injury, death or loss of property. It is restricted to industries that deal in hazardous substances. There is an upper limit of Rs 50 crore that can be claimed as compensation from the insurance companies.

The National Environments Tribunal Act, 1995, fixes liability and provides compensation for cases on environmental damage.

MSIHCR, 1989, attempts to prevent accidents in chemical units and ensure that the culprits and local organisations are primed.

The requirements include: preparation and regular updating of on-side/off-site plans; potential victims to be informed of the plans and mock drills to be conducted every half year for on-site situations and yearly for off-site incidents.

The Chemical Accidents (Emergency Preparedness, Planning and Response) Rule, 1996, complements the set of rules above. A more fortified crisis management structure is envisaged for the handling of industrial accidents at the Central, state, district, and local levels. The responsibility of constituting these groups lies with the Central Government, chief secretaries of the states, and the district collectors for each level.

The new statutory amendment to the Factories Act creates a general duty for occupiers of hazardous installations to ensure the health and safety of workers, adequate safety systems, equipment arrangements and maintenance, training and information, safe practices and processes.

The Environment Protection Act provides for better zoning of industrial locations, a more intensive system of inspections, and the development of prevention guidelines and emergency response systems.

An Emergency Preparedness Plan (EPP) has to be prepared both by the occupiers and the district authorities.

The District Industrial Health and Safety committee, under the District Collector, is supposed to supervise the safety aspect in hazardous industries and the implementation of precautionary measures. A crack committee is supposed to supervise the most hazardous industries.

The EPPRCA focuses on the management of chemical accident emergencies. The key requirements are the formation of a crisis alert system, and crisis groups at state, district and local levels.

In 1992, India made an environmental audit report mandatory for all industries covered by the Air and Water Pollution Acts and the Hazardous Wastes Handling Act [5]. The reports are to be submitted to the State Pollution Control Boards.

In the same year, the Indian Government effectively ratified its participation in the Basel Convention on the Trans Boundary Movement of Hazardous Wastes [5].

The Freedom of Information Act in Madhya Pradesh, January 31, 2003, gives every citizen the right to inspect or obtain copies of records and documents of any government department, office or institution in public interest, which have been prepared in the previous three calendar years. This includes companies and corporations where the State Government of MP holds 51 percent equity. However, the list of restricted information is long (*Hindustan Times*, Feb 5, 2003).

The Department of Environment (DoE) and the Controller of Explosives & Mines (CEM) hold common charge of safety standards but with widely differing scope [6]. The Pollution Control Boards (PCBs) stick to issues of environment, and the factory inspectorate with workers. The storage of hazardous gases under pressure is the responsibility of the CEM.

An amendment in 1986 (section 7B) stated that the supplier of a foreign technology should not be held liable for the technology supplied by it [6]. No foreign collaborations are now sealed without this escape hatch.

The Factories Act, 1948, provides the broad framework for worker protection in factories against accidents and occupational diseases. The states define and implement the laws [6]. The Employees' State Insurance Corporation Act provides social security as well as compensation for diseases and accidents caused by work. The Workmen's Compensation Act, 1923, seeks to protect workers who are not covered under the aforementioned Act.

8.7.1.2 The Disaster Management Institute

The Disaster Management Institute in Bhopal (DMI) was established in 1987, after the Bhopal Gas Disaster, by the Government of Madhya Pradesh [7]. It is the only institute of its kind in India. Its aims are to provide training in disaster management, to carry out research-oriented studies concerning causes and effects of disasters and their prevention and mitigation by management, to collect information concerning hazards and disasters, and to offer consultant services to industries and others.

A State Crisis Group has been set up, and action plans for flood disaster preparedness and earthquake disaster management have been prepared. A project on reducing the risks of environmental deterioration due to pollution from chemical industries was started in 1994 in co-operation with NORAD (Norwegian Aid and Development Agency).

Training courses are organised on prevention and mitigation measures in the event of fire, bomb threat, earthquake or release of toxic agents. After September 11, 2001, the programme has been expanded by including nuclear, biological, and chemical release (*Hindustan Times*, Jan 28, 2003).

8.7.1.3 The Environment Protection Agency

The Environment Protection Agency came into being in 1996. It should lead to a new framework of legislation to ensure that the Bhopal disaster is not repeated.

8.7.1.4 Other authorities

The Central Pollution Control Board (CPCB) was established in 1988. Its responsibilities include coordination of the functioning of the state boards, providing technical assistance and guidance, initiating investigation and research, planning and organising training of staff, collecting and publishing technical data, etc.

State Control Boards exist in every state.

8.7.1.5 Non-governmental organisations

Given the environmental effects of the Indian development policy, the growth of an environmental movement was inevitable [5]. Population response to pollution damage began as early as 1958. Following the massacre at Bhopal, opposition to toxic production strengthened.

NGOs in India have found it necessary to co-operate in networks that together cover a large area with broad competence.

The Other Media and the Delhi Science Forum together with the survivors' organisations are responsible for keeping the criminal cases alive. The Bhopal Group for Information and Action (BGIA) supports local survivors' organisations and keeps in touch with an international network. The Sambhavna Trust not only provides health care, but also different kinds of support to survivors and other documentation.

8.7.2 The international community

8.7.2.1 International labour organisations

The International Labour Organisation (ILO) has developed a series of conventions:

- The Code of Safety, Health and Working Conditions in the Transfer of Technology to Developing Countries emphasises the continuing responsibility of all parties in hazard management, specifying detailed procedures for the design, operation and alteration of hazardous technologies.
- Convention No. 170 of 1990 deals with the safe use of chemicals at work.
- Convention No. 174 of 1993 deals with major industrial accidents.
- The Tripartite Declaration of Principles concerning Multinational Enterprises and Social Policy stipulates that a national or multinational enterprise with more than one establishment should provide safety measures without discrimination to the workers in all its establishments, regardless of the place or country in which they are situated.

An ILO-aided project has identified nearly 600 major hazardous installations that will be included in a developing control and inspection plan [2].

The International Confederation of Free Trade Unions (ICFTU) is represented by the United Nations Commission on Sustainable Development (CSD), which is the UN body meant to co-ordinate Agenda 21 efforts. For the 1998 session, a declaration was prepared by the ICFTU and the Trade Union Advisory Committee to the OECD (TUAC) [8].

Trade unions also participated in the preparatory work for the Third European Ministerial Conference on Environment and Health, London 1999. The declaration includes a paragraph on health, environment and safety management.

Local trade unions are involved in environmental issues. In India, the West Bengal Cha Mazdoor Sabha is pressing for amendments to the Plantations Labour Act to bring protection and training to team workers exposed to agro-chemical hazards [8].

8.7.2.2 Inter-state organisations

Inter-state organisations have also drawn up declarations and conventions [2]:
- The *Bhopal Resolution* of the European Parliament calls upon European firms to maintain levels of safety abroad that are comparable with those in place in their home operations.
- In the *OECD Guidelines on Multinational Enterprises*, the member states agree to regulate their multinationals doing business abroad, to ensure that their operations "are in harmony with national policies of the countries in which they operate".
- The *OECD Code on Accidents Involving Hazardous Substances* emphasises the imperative of providing citizens with full information and enhancing their role in the decision-making process.
- The 1982 proposal *UN International Code of Conduct on the Transfer of Technology* articulates the responsibilities of the contracting parties.
- The 2003 proposal *UN Human Rights Norms for Business: Towards legal accountability* describes the responsibility of the states as well as of the trans-national corporations.

8.7.2.3 The international non-governmental community

After the Bhopal Gas Leak, several national NGOs, who also co-operated in networks, adopted the issue. These networks are still functioning, although the work may have turned to toxic chemicals in general, with Bhopal as an example. However, these networks have been engaged in collecting money for the Sambhavna Trust.

The international non-governmental community has developed its own declarations. The Permanent Peoples' Tribunal has developed a charter on Industrial Hazards and Human

Rights, following a row of international conventions and guided by other declarations [9]. Some points are:

• Right to organise;
• Right to appropriate health care;
• Right to a living environment free from hazards;
• Right to environmental information;
• Right to enforcement of environmental laws;
• Right to relief and compensation.

At the 5th and 6th World Conferences on Injury Prevention and Control (New Delhi 2000 and Montreal 2002), a Declaration on People's Right to Safety has been produced [10]. The aim is that right to safety should be endorsed by the United Nations. The declaration includes:

• Right to participation in safety promotion;
• Redress and avoidance of risk;
• Environmental Monitoring;
• Emergency Preparedness;
• Relief and compensation;
• Right to fair procedure;
• State responsibilities.

8.7.3 Implementation

Data is gathered by each state and collected by the Director General Factory Advice Services and Labour Institute (DGFASLI). Only 9 of the 35 states have the entire set of relevant information with less than 15% of the work force included [6].

A government survey in the aftermath of the Bhopal catastrophe identified over 7,000 potentially hazardous plants in the country [5]. In 1995, it was reported that of some 9,000 industrial units screened by the Delhi Pollution Control Committee, only 283 had installed effluent treatment plants. As much as 70% of India's water resources are polluted.

The Ministry of Environment and Forest (MoEF) has identified 1,254 "major accident hazards" (MAH) units and 12,702 "ordinary" hazard units [6].

The industries' environmental audit reports are to be submitted to the State Pollution Control Boards. However, these Boards are overloaded and understaffed [5]. It has been pointed out that the Water Act and its subsequent amendments have been largely ineffective, mainly because of lack of resources. The laws rely on voluntary compliance and leave no scope for litigation. For polluting firms, the cost of compliance has exceeded the cost of non-compliance.

According to a 1992 report of the Comptroller and Auditor General, The Central Pollution Control Board (CPCB) is not vested with the powers to discharge its responsibilities, as the state boards are not required to consult the CPCB nor to comply with the standards that have been set [5]. The CPCB cannot take over the activities of the state board if the latter defaults. The ministry notifies standards in consultation with the state boards, and not the central authority. The CPCB is also understaffed. The state and central governments can blame each other for the continuing failure of environmental regulations.

The State Control Boards are generally understaffed, with few or no experts and very little resources. It is likely that the state pollution control board officials are susceptible to bribes, as they are not very well paid.

Another problem relating to industrial structure is that it includes many small firms (both unorganised and household units) that lack adequate capital resources, physical space and the skills needed for designing pollution systems [5].

In 1995, it was discovered that the Indian recycling industry was actually dumping large quantities of hazardous waste (after recovery) imported from abroad [5]. The Government of India did not know how much of hazardous waste was being generated within the country and how it was being disposed of. The Supreme Court set up a committee to examine all aspects relating to the disposal of hazardous wastes in India and to make recommendations. In the report 2001, it was stated that the international waste trade continued to import hazardous wastes into India despite the directions of the Supreme Court banning such import.

There is a gross negligence of the occupational health care system [5]. In Gujarat, the Chief Inspector of Factories had not been notified about the incidence of any of the 29 notifiable diseases under the Factories Act, 1948, though 34% of its population are industrial workers. The Tata Memorial Centre in Mumbai states: "No reliable information is available on occupational cancer in India". Occupational tests carried out by government agencies or labour organisations are sometimes not accessible because they are considered the property of the client (factory operator).

From unofficial sources, estimates have been made that for every 100,000 workers, as many as 7,000 may be injured every year, while 75 die [5]. For a total workforce of 20 crores that means about 150,000 deaths are recorded every year.

Another problem is that a large part of the workforce is not employed and often not even recorded as employees. In the tanning industry, only 35–40 % of the total workforce is covered under any labour welfare scheme [5]. The most disadvantaged workers are women.

Examples of state repression towards environmental activists are legion, including the use of the Official Secrets Act and the murder charges against activists in Bhopal. The Terrorist and Disruptive Activities Act has been used for arresting the top leadership of the anti-nylon movement in Goa 1994 [5].

In Thane, in 1993, residents died or were hospitalised after exposure to toxic fumes from a viscose filament plant [5]. A team from the State Pollution Control Board visited the site. Citing gross negligence by the factory managers, the Board ordered the company to close down the factory. However, the company managed to mobilize the trade unions to argue that closure would cause unemployment, and the factory continued to function defiantly.

It is said that while one hand offered bribes, the other threatened [5]. When prosecuted, private companies sometimes threaten to shut down the factory rather than install pollution measures. Then the Government immediately relaxes the pressure because it needs the industry for economic expansion.

The Government is still keen on establishing chemical industries. There are several examples of how decisions are hastened, without considering either opposition or documented expected environmental effects [5].

The companies have learnt from Bhopal. The following sentence was included in an agreement: Du Pont and its representatives will be "harmless from any claims made in the Republic of India against representatives of Du Pont or its assignees alleging bodily harm or death sustained as a direct result of, or in direct connection with, the performance of this Agreement".

A spokesman for the CITU (Centre for India Trade Unions) alleges that accidents every year result in over 30,000 deaths and more than 200,000 injuries, while over a million workers are affected by occupational hazards.

Thousands of cases have been filed against violating organisations and individuals [5]. But while there have been some victories; overall, the legal struggle has shown great limitations. While there have been innovatory judgements and judicial activism, it is fair to say that the legal system has provided limited redress.

8.7.4 Comments

We do not yet know whether these laws, regulations, conventions and declarations will be enough to prevent another Bhopal disaster. As long as governments do not implement the appropriate laws and regulations, nothing will happen. One type of convention is still missing: the multinational corporations' code of ethics.

Srishti [4] points out that the Indian laws and rules focus on how to manage the toxicity and the waste. The processes that lead to the hazards are not examined. The laws have provided a sense of security, and, thereby, have legitimised the continuance of a toxic legacy.

India is called "the world's largest democracy". India is also the country of Gandhi, the man who led India to freedom through non-violence. The tradition is strong, demonstrations and sit-ins are common and accepted instruments to influence politics and decisions. This makes it all the more remarkable that state violence and repression are so commonly accepted by the authorities and politicians, and so commonly used against the people themselves.

Seven trans-national companies are testing the *UN Human Rights Norms for Business* in their own activities (*Dagens Nyheter,*

Jan 19, 2004). They discuss who should monitor the companies. It is not uncommon that governments break the rules on human rights. One of the testing companies is the Swedish ABB, which has been involved in the building of big dams and other constructions that threaten human rights as well as the environment.

The ILO conventions are not an assurance that the labour organisations will co-operate in order to reduce pollution. The catastrophe in 1997 at the Halland Hill in Sweden, where toxic chemicals used for tightening a tunnel leaked out into the groundwater, would not have occurred if the French trade union had informed the Swedish trade union of the nature of the compound.

There is often a conflict between the workers' need for employment and a society's need to reduce pollution. This conflict seems to have found a solution when trade unions start to co-operate with employers and governments to make the plants more environmentally friendly.

However, the NGOs inside India will keep an eye on how the Government of India fulfills its plans. The international networks of NGOs will go on lobbying, demonstrating, writing articles etc., to influence the development of the world in an environmentally friendly direction.

It is to be noted that environmental awareness among the people and the government is now being cited as a major factor in making India a less appealing place for foreign investors [5]

8.7.5 References

1. Sengupta, M., *Emergency Preparedness Plan for Chemical Hazards*, in *Refresher Course for Top Executives. Management of Chemical Accidents.* 1994, Disaster Management Institute: Bhopal.
2. Cassels, J., *The Uncertain Promise of Law: Lessons from Bhopal.* 1993, Toronto: University of Toronto Press Inc.
3. Chauhan, P. S., *Bhopal Tragedy. Socio-legal Implications.* 1996, Rawat Publications: Jaipur.
4. Srishti, *Surviving Bhopal. Toxic Present-Toxic Future. A Report on Human and Environmental Chemical Contamination Around the Bhopal Disaster Site.* 2002, The Other Media: Delhi.

5. MacSheoin, T., *Asphyxiating Asia*. 2003, Mapusa, Goa: Other India Press.

6. Meenakshi, N., *Industrial Disasters Working Towards Oblivion*, in *India Disasters Report. Towards a Policy Initiative*, S. Parasuraman and P. V. Unnikrishnan, Editors. 2000, Oxford University Press: New Delhi.

7. *Information Brochure*. 1994, Disaster Management Institute: Bhopal.

8. *Commitments for Sustainable Development. Trade Unions at the Commission for Sustainable Development. Special "Business and Industry" Segment*. 1998, ICFTU: Brussels.

9. *Charter on Industrial Hazards and Human Rights. Permanent Peoples' Tribunal*. 1996, PesticideTrust: London.

10. 6th World Conference on Injury Prevention and Control, M., Canada, 15th May 2002, *The Montreal Declaration on People's Right to Safety*. 2002, http://www.iitd.ac.in/tripp/rightto-safety/Montreal%20declaration%2015-05-02.html.

8.8 The Roles of the Actors

8.8.1 The role of Union Carbide

Union Carbide Corporation (UCC) was one of the owners of Union Carbide India Limited (UCIL). UCC owned a little over half of the shares and still denies any criminal liability.

Cassels [1] discusses the relationship between multinational corporations and their host countries in the developing world. "The terms and conditions of their relationship are fundamentally determined by the interests and abilities of the parties, which include the relative poverty and developmental needs of the host state, balanced against the power and economic imperatives of multinational business organisations. In this dynamic, workers and communities are rarely considered to be parties."

The MIC plant is falling apart (Christian Saltas)

The MIC plant is falling apart (Ingrid Eckerman)

UCIL's personnel management policy led to many skilled workers leaving the plant. Management warned workers against taking part in union activities. It formed a safety committee that included workers, but the committee only had a formal function. Agitations regarding safety were stopped by instituting a commission of enquiry [2]. When the workers' union sent letters about hazards and pollution to the managers of the plant and the factory inspector, they received no reply [3].

The statements by UCC on their responsibility for the Bhopal plant are somewhat contradictory. After the 1981 accident in Bhopal in which a worker was killed, a telex said that improvements "will receive close attention by the management committee in New York" and that it was "very essential" that this committee know the "specific actions" taken to prevent recurrence. Another memo said, "No design changes have been made without the concurrence of general engineering or Institute plant engineering", referring to Carbide's corporate engineers in Institute West Virginia [4]. After the leakage, however, UCC stated that UCIL was an independent affiliate, and that UCIL had the full responsibility for management, including safety measures.

The vent gas scrubber today (Christian Saltas)

The Permanent Peoples' Tribunal in 1992, having received oral and written testimonies as well as having studied Carbide's own documents, concluded that the "Union Carbide Corporation, its subsidiary, Union Carbide (India) Ltd., and key officials of both are clearly guilty of having caused the world's worst industrial disaster" [5].

Many observers maintain that UCC treats information on MIC like trade secrets [1, 4]. Immediately after the leakage, the director of MP Health Service had asked for information of the composition of the cloud. He got none at all [6]. It is generally assumed that UCC has much more information on MIC than they ever came close to releasing. The video produced by UCC talks a lot about very thorough research, but there is no information about the results of such research [7].

Mr Warren Anderson, chairman of the UCIL plant in Bhopal, arrived in Bhopal a few days after the disaster. He wanted to see for himself that everything that could be done was being done. He was immediately put under arrest by the Government of Madhya Pradesh, but was subsequently released [6].

The compensation issue is very important. The final decision was postponed for five years and the amount was much lower than expected – actually, only the insurance sum plus interest. If the leakage had happened in the US, UCC would have had to pay many times this sum.

Morehouse [8] maintains "that outlasting the victims" was a key element in the Carbide strategy. The sabotage theory turned UCC from a victimiser to a victim.

After the leakage, UCC provided expensive equipment to hospitals and funded the Red Cross clinics and the Eye Hospital. This led to further suspicions. Warren Anderson, the manager of the Bhopal plant, was on the board of the Red Cross in America. The chairman of the local Red Cross unit was responsible for the security of Carbide plants in India. Local activists alleged that medical records from the Red Cross clinics were sent to Union Carbide [4]. It is also said that gas victims were attracted to the eye hospital by the free rations and edible oils offered, but had to undergo medical tests in return [4]. After some years, the support was withdrawn, and the Red Cross clinics had to close (see 8.4.1.4).

The Bhopal Hospital Trust (BHT) (see 8.4.1.2) is said to be a way for UCC to get out of India. The Government of India permitted the corporation to sell its shares in UCIL on condition that they placed the money in the BHT. Despite protests, the main part of the fund has been used for a highly specialised hospital far away from the affected area, instead of for a network of primary health clinics.

UCC has conducted a major public relations campaign claiming moral responsibility for the disaster, to give the picture of a company, which does everything to compensate for its mistakes [8]. Meanwhile, in the courts, they denied all responsibility for what happened at the Bhopal plant.

In December 1985, the year after the leakage, Union Carbide announced a major restructuring programme [9]. A large part of the capital base was sold off, major assets were divested and the proceeds distributed to the shareholders. This reduced the assets from which to pay eventual Bhopal claims.

If the company management had made proper offers to help on a scale appropriate to the magnitude of the disaster, they might have been confronted with legal suits by shareholders for mishandling company funds [8]. Carbide's stock price rose US$ 2,00 a share on the day the February 1989 settlement was announced.

When information on the catastrophe spread, a psychological shockwave hit the employees of UCC all over the world [6]. They lost confidence in "Carbide's strong corporate identity". On December 6, 110,000 employees at the 700 factories and laboratories stopped work for ten minutes. Lapierre considers this is the reason why it was so important for UCC to show that the company was not guilty, and that sabotage was the cause.

Many attempts to contact UCC have been made through the years. The author of this book sent the draft of her essay to UCC and asked for comments, but got none. The management failed to respond to any requests for interviews and information from Lapierre and Moro [6]. Rhône-Poulenc and Dow Chemicals were more positive.

Union Carbide India Limited (UCIL) changed its name to Eveready Industries India Limited. The Bhopal leak was the start of a decade of decline for UCC. The agricultural division was sold to the French company Rhône-Poulenc, the company that produced the packing material that poisoned the groundwater at the hill in Halland in Sweden [10]. In 2001, the merge with Dow Chemicals was finally effected. Dow has a similar history of neglecting environmental and human safety as UCC. This is the company that produced the defoliant Agent Orange that was used during the Vietnam war. Dow became one of the largest – or even the largest – chemical companies in the world.

The survivors now claim that Dow Chemicals has taken over the responsibility for taking the necessary measures. The words uttered by Dow's president, however, show the transnational's attitude: "Clearly we're enormously aware of Bhopal and the fact that particular incident is associated with Union Carbide, [but Union Carbide has] done what it needs to do to pursue the correct environment, health, and safety programs. – It is not in my power to take responsibility for an event which happened fifteen years ago, with a product we never developed, at a location where we never operated" [11].

8.8.2 The role of the governments of India and Madhya Pradesh

The Government of India was the other owner of UCIL. The ownership was divided among different financial institutions. The Union of India had sufficient bargaining clout to ensure that the construction of the plant and of much of the necessary equipment was undertaken in India by Indian firms [1].

Carbide also claims that it was pressured into manufacturing MIC by the Government of India because of the latter's policy of urging foreign companies to "indigenise" their production [8].

The city of Bhopal did have a development plan and zoning regulations [1, 12]. The UCIL plant was classified as "general" in order not to have to relocate to the industrial area. When the municipal planning administrator issued a notice to UCIL to relocate the plant, he himself was relocated. Sixteen other industrial facilities were relocated [1]. In 1982, it was the labour minister who objected to relocation of the plant. Settlements were allowed to grow in the immediate vicinity of the plant, and were authorised by the government in 1984.

Workplace safety and health conditions are governed by the Factory Act, 1948. Legislation is federal, but inspection of health and safety conditions is the responsibility of the Government of Madhya Pradesh. It was not until 1981 that India enacted the Air Pollution Act, and not until 1986 that it enacted comprehensive environmental protection legislation [1].

At the time of the disaster, the state of Madhya Pradesh had only 15 factory inspectors.

The factory inspectorate received complaints from the union leaders, and they conducted inspections in 1981 and 1982. No action was taken by the government to ensure that the hazardous conditions in the plant were corrected.

Atmospheric tests within the UC Bhopal plant conducted by the MP Environmental Pollution Control Board (EPCB) indicated that ISI limits of the biological oxygen demand and chemical oxygen demand levels were crossed 10 to 100 times. According to the Air Pollution Act, the government had the authority to revoke the plant's licence, but no action was taken. The Pollution Control Board spokesman admitted that there was no consistent monitoring of the effects of the plant on the air and the water around the place, not even when animals which had drunk water from the pond died [13].

In 1982, The Employment Minister of Madhya Pradesh had announced in the assembly rostrum: "There is no cause for concern about the presence of the Carbide factory because the phosgene it produces is not a toxic gas." The journalist Rajkumar Keswani had made personal investigations. He sent a letter with the findings to this minister as well as to the Chief Justice of the Supreme Court. He got no answer [6].

The steps taken by the Chief Minister, before and after the leakage, seemed to have the overall aim of helping him to win the elections [6].

After the leakage, India's Central Bureau of Investigation (CBI) took over the inquiry. The UCC people who had arrived from USA were denied access to the plant. The factory's archives were moved to a secret place.

It was not only Union Carbide, but also the Indian authorities who did not release information and who tried to cover up what actually happened:

- Voluntary organisations were not allowed to enter government relief camps [4].
- The "cause of illness" on hospital case papers was changed from "MIC poisoning" to something else [4].
- Post-mortem results and case histories of gas victims were declared as classified [4].
- The results of the Tata survey in 1984 have still not been released.
- The reports from ICMR were not released until ten years after the leak.
- The ICMR issued strict instructions to all doctors, both private and government ones, not to disclose their findings to the press or the public [4].

- The sealing of the plant by the CBI effectively sealed information about the plant and what really happened there. Reporters were denied access to the plant by the Bureau, as was the international trade-union mission in March 1985 [4].
- When press reporters contacted the Environmental Planning and Co-ordination Organisation (EPCO), the officials were tight-lipped [14].
- The Indian government proposed destroying the remaining MIC in tank 610, thus making its analysis and testing impossible [4].

Jones [4] discusses the significance of "Operation Faith" (see 7.3.7) to re-establish social control by breaking up the communities and scattering people from Bhopal throughout the state of Madhya Pradesh.

State repression towards the opposition in Bhopal has been legion [4]. On May 19, 1985, a large number of victims, mostly women, were beaten up by the police during a demonstration to press for adequate relief and medical facilities [15]. On June 25, 1985, there was a "midnight swoop" which resulted in the arrest of around 40 activists, including doctors, and the closure of JSK (the People's Health Centre). The aim was to prevent a protest rally on the following day, during which 31 activists were arrested. The same month, the police closed a women's centre providing food and housing a library.

A scientist contended that there was a strong pro-Union Carbide lobby within Indian government circles, at both the national and state levels, which had manufactured what he called "a string of lies" to conceal information about the number of deaths, the nature of the gas, and the rehabilitation and relief efforts [15].

The Government of Madhya Pradesh has spent a great deal of money on economic aid, vocational training, environmental improvement and medical rehabilitation [16, 17]. Unfortunately, no assessment of the investments has been done, which leaves the field open for guesswork and suspicion. Money meant for improving the living conditions of the survivors was spent on routine municipal activities in areas that were not or only slightly affected by the gases.

The support or lack of support to the survivors has, to a great extent, been dependent on politics. One example is the

connection between distribution of ration cards and the elections (see 8.3.3.2). Local political activists associated with the
Congress (I) party carried out the June 1985 state repression.
The demolition of 752 houses belonging to Muslims (see 8.3.5)
took place during the rule of the BJP (the fundamentalist Hindu
party).

In a report on the construction of hospitals from the Comptroller and Auditor General, remarks were made about high administrative costs, delays and increasing costs, and "excess
costs" because of incorrect calculations.

In 1998, the Government of MP decided to construct a memorial at the site of the UCIL plant.

The Permanent Peoples' Tribunal in 1992 concluded that
"the Government of India and the Government of Madhya
Pradesh are also clearly guilty of violating the rights of the victims, not only under the international human rights law, but
also under the Indian Constitution" [5][5].

Today, the Madhya Pradesh (MP) State Government and the
Rehabilitation Centre are fighting on the remaining money
(about Rs 1,360 crore) [18]. MP wants money to provide the
city with potable water from the Narmada river. The Rehabilitation Centre, that is in charge of the money, is not approving
this.

Members of IMCB have, for years, asked for governmental
support for arranging a medical conference for Bhopal's doctors.
In November 2003, the response from the Chief Minister was
rather positive. But after the elections, which the BJP party won,
a letter arrived, saying that "the Government of Madhya Pradesh
would not be in the position to co-host an international conference on Bhopal gas victims at Bhopal during 2004" [19].

8.8.3 The role of the medical and scientific societies

There are many examples where the behaviour of the physician
or scientist can be questioned from an ethical point of view [4,
6, 20, 21].

• The chief medical officer at the plant stated that "the safety
  precautions we took were the best possible. In fact we used
  to think that we were overdoing the safety". When Hamidia

hospital was overcrowded with dying patients, he maintained that MIC is not a deadly gas, just irritating. He recommended that the patients should drink plenty of water and rinse their eyes.

- The senior regional adviser for the WHO, said that survivors faced no risk of paralysis or kidney and liver diseases, and that pregnant women and foetuses would suffer no damage.
- The medical and scientific actions over the sodium thiosulphate issue are described in section 7.4.5. It is said that the key medical officials in governmental institutions, who took a standpoint similar to the UCC doctors, had ties to UCIL even before the leakage [15].
- The medical establishment was not interested in warning pregnant women of the risk of the teratogenic effects of the gases [4].
- In interviews, doctors tended to give the impression that the survivors were exaggerating their symptoms to get relief, and that they did not want to work [22].
- Directly after the leakage, Indian scientists were quoted in the newspapers, reassuring the public that the patients would not suffer permanent injuries. One ophthalmologist first admitted that late effects could be very serious, but later withdrew this.
- In 1985, an American ophthalmologist sought to underscore the transient effect of MIC on the eyes as well as on the lungs and minimise its permanent damage at a presentation for an NGO conference [15].
- The reports of the three American specialists, sent to Bhopal by UCC, have not been released.
- A team of Bhopal doctors was invited by UCC to go to the US to make a depositione in American courts that there was a high incidence of tuberculosis in Bhopal, which would explain the lung damage. The headline of an article describing this was "Carbide buys up Bhopal doctors" [15].

Within the first week of the disaster, several "medical experts" visited Bhopal, sponsored by UCC. While their stated purpose was to help people in distress, it did not take long before their real mission became clear. In their meetings with senior doctors as well as in their interviews with the media, these experts repeatedly emphasised that the leaked gases would not have any long-term health effect on the exposed persons [20].

- The behaviour of scientists involved in epidemiological and clinical research is described in chapter 8.2. It is said that some of the scientists had ties to UCIL and the government.
- Today, the medical society in Bhopal does not show much interest in the gas victims. They prefer to support super-specialised hospitals rather than primary health care. There is no interest in developing adequate treatment, nor in doing research. Only a few doctors have shown any interest in Sambhavna.
- A group of American doctors visited Bhopal after the leakage, on behalf of Union Carbide. Their role was unclear and it was suspected that they were sent to perform "a medical intelligence-gathering exercise for Carbide's legal defence". In addition, their message was that "the gases could not have any long-term health effects".

But the medical profession has also made very solid contributions:
- Prof. Heyndrickx of the University of Ghent toxicology department, Belgium, said on December 5, 1984, that hundreds of people would die of secondary respiratory and neurological effects, while "it is certain that unborn children will be born with enormous deformities. One cannot be sure about children conceived later by survivors, but there is a significant risk".
- Many of the volunteers who came to Bhopal were physicians, who did very good work during the first chaotic weeks as well as later. Many of them have continued to keep themselves informed about the Bhopal issue and provide support when needed, although they are no longer actively working for the survivors.
- There are still physicians from abroad engaged in Bhopal, for example as a follow up of the International Medical Commission on Bhopal 1994, and as volunteers at Sambhavna clinic.
- At Sambhavna, there has been high demand for the "right spirit" of the doctors recruited, although it has not always been easy to find them.

8.8.4 The role of the judiciary

Details on the legal part of the saga is described by Jayaprakash [23].

A number of serious offences were registered on December 3, 1984, at the local police station [24].

On December 7, Warren Anderson, chairman of UCC, arrived at Bhopal. The Indian Foreign Affairs minister had promised the American State Department that nothing would impede Anderson's journey. He had also been promised a meeting with Rajiv Gandhi, the then Prime Minister of India, and the Chief Minister of Madhya Pradesh. In spite of this, he was arrested immediately. On an order from the Chief Minister, he was accused of "culpable homicide causing death by negligence". Some hours later, he was released on bail of Rs 25,000 (about 2,000 dollars). When asked if he was prepared to come back to India to answer any legal charges, he said he was [6]. Warren Anderson has never appeared in the court despite repeated summons. During long periods, he has succeeded in hiding from Indian authorities and activists.

A High Court action forced the preservation of 15 kg of MIC from tank 610, when the Government of India wanted it to be destroyed [4].

In 1987, the Central Bureau of Investigation of India (CBI) brought criminal charges in the court of the CJM, Bhopal. The accused are Warren Anderson and six other persons, as well as the UCC in the USA, Union Carbide of Hong Kong and Union Carbide at Kolkata (Calcutta) [24–26]. They are accused of culpable homicide causing death and of poisoning animals.

The Supreme Court of India accepted the argument that because the government had taken over the victims' rights, it should also, as part of the social contract, provide for their welfare in the interim [1].

In the settlement of 1989, the Supreme Court said that "all criminal proceedings related to and arising out of the disaster shall stand quashed wherever these may be pending" [25]. In 1991, the same court, but another judge, decided that the quashing of the criminal proceedings was not justified and directed that the criminal proceedings should continue.

In 1996, the Supreme Court reduced the charges from "culpable homicide" to "criminal negligence", which reduces the maximum sentence from 10 to two years and provides an opportunity for the case being dropped [27]. Since UCC had deregistered UCE, Hong Kong, in 1992, the CBI has expressed its inability to proceed against it [24]. In 2002, the CJM's court reaffirmed that Warren Anderson stood accused of "culpable homicide". It directed the Indian Government to secure the extradition of Anderson and Dow-Carbide representatives to face trial in India

The criminal case is still going on. The Indian accused are currently appearing before the District Court, but at an extremely slow pace of less than one hearing per month. In 1999, the Indian Government had still not taken any steps towards seeking the extradition of Anderson [24]. The whereabouts of Anderson have been unknown for long periods and representatives for the accused often do not appear at the sessions. In 2002, Greenpeace discovered him in New York. In December 2002, the prosecuting council sent papers for his extradition from the US to the Ministry of External Affairs in India.

The Court has permitted three survivors' and activists' organisations to assist the prosecution during the trial [28]. Without the engagement of activists' and survivors' organisations, the case might have ground to a halt years ago.

Directly after the catastrophe, hundreds of American lawyers invaded Bhopal. They approached survivors and asked them to put their fingerprints on a power of attorney letter so that they could represent Bhopal survivors in the American courts. The lawyers realised that there might be big money to earn. Various suits for compensation were filed in various District courts in USA. One such suit was for 15 billion US dollars [25]. In 1985, all such suits were assigned to the District Court of the Southern District of New York, where Judge Keenan was presiding.

The civil proceedings concerning economic compensation are described in chapter 8.3.3.1. There is no explanation as to why the Union of India in February 1989 so suddenly agreed to the compensation sum of US$ 350 million. The previous day, however, the Central Bureau of Investigation had been given permission by the American government to inspect the Union

Carbide plant in West Virginia. This inspection was later not permitted.

Union Carbide has employed groups of lawyers in the US as well as in India. The strategy has been to postpone any legal judgement, and thus to delay the provision of relief and justice to the victims [8].

Networks of Indian lawyers have involved themselves on behalf of the survivors. They have followed the different legal turns, debated and fought for the rights of the survivors. The Lawyers Collective in Mumbai is one of these groups [29].

In November 1997, the Human Rights Commission of Madhya Pradesh registered a case of human rights violation against officials of the state government [28]. This was in response to a complaint filed by BGPMUS and BGIA on the denial of medical care and on the negligence in monitoring exposure-related morbidity and mortality.

Seven individuals and five organisations filed a class action suit in 1999, in the Federal District of New York, against UCC and its former chairman Warren Anderson [26]. In 2000, Judge Keenan of the Federal Court of the Southern District of New York dismissed the suit on the grounds that the Bhopal Act (see 8.3.3.1) prevented individuals or organisations outside the Government of India from bringing an action against UCC or its officials. An appeal was filed before the Second Circuit Court of Appeals.

In 2000, a class action lawsuit was filed against Dow Chemicals in the US District Court for the southern District of New York [26].

The judge of the Labour court in Bhopal in 2000 directed UCIL/Eveready to pay full wages to those workers, who were sacked at the closure of the factory in 1985, till such time as their services were terminated in accordance with the law [26]. This would amount to about Rs 1 million per worker. However, the company has filed an appeal against the judgement.

8.8.5 The role of survivors

The day after the leak, the survivors stormed the Bhopal plant (see 7.3.3.). In January 1985, large-scale protests were organised against the withdrawal of the free distribution of rations [4]. This resulted in police violence and arrests. After this, the rations were restored.

Identification with being a "victim" or a "survivor" is still strong. On the walls, the call "UC out of India" is still evident. Simultaneously, the black cat, the logo for batteries produced by Eveready, a UCC subsidiary, is visible in several places, and the public does not seem to react.

Over the years, many survivors' organisations have been founded and have disappeared (appendix 4).

In 1994, there were four survivors' organisations which continued to be active [30]. They had marched on over 400 occasions in non-violent demonstrations and policemen had attacked them at least 40 times. They had participated in "dharna" (sit-ins) on at least 300 occasions, often combined with hunger strikes. The organisations had different bases: one was for pensioners, another was a trade union for women at a former workshed, one organised thousands of women and activities in politics, and the fourth organised and educated women at a grass roots level.

Currently (2004) four organisations are active. *Bhopal Gas Peedit Mahila-Stationery Karmachari Sangh* is a trade union for the only remaining stationery workshed. It has around 75 members, and the leaders are women (one Hindu and one Muslim). They have many supporters who are not union members but join in the demonstration and rallies. *Bhopal Gas Peedit Mahila Udyog Sangathan (BGPMUS)* organised several thousands of women, who chose a man as their leader. *Bhopal Gas Peedit Mahila Purush Sangharsh Morcha* (Bhopal Gas Affected Women's and Men's Struggle Front) is a breakaway group from BGPMUS and is working closely with the stationery women. *Gas Peedit Nirashrit Pension Bhogi Sangharsh Morcha, Bhopal (GPNPBSMB)* pays special attention to aged survivors, widows and orphans.

The actions of the survivors' organisations have contributed to the payment of interim and final compensation, to the criminal case not being closed, and have ensured that the Bhopal gas tragedy is not forgotten, either in India or internationally.

On the day of every anniversary, there is a large rally with several thousand participants, men and women. At the front of the rally is a very large effigy of Mr Warren Anderson. The rally

The statue of the fleeing mother is the symbol for the victims (Christian Saltas)

Women are active in the fight for justice (Ingrid Eckerman)

The largest effigy at
December 3 (Ingrid Eckerman)

Piece of art for the 10th anniversary
Artist unknown.

Rally at December 3 (Ingrid Eckerman)

walks along the old city, makes stops at different churchyards and passes the plant. Finally, in the evening, the effigy is burnt.

The survivors' organisations are very active in The International Campaign for Justice in Bhopal (see 8.8.5.3). In January 2004, 350 survivors went to Mumbai (Bombay) for the World Social Forum.

8.8.6 The role of international organisations

The role of the WHO is unclear. Only in the Trade Union Report is there some information [31] about it[15]. A WHO representative, Dr Jaeger, arrived on December 8. However, the WHO does not seem to have played any important role in the post-event phase.

In Madhya Pradesh, the Norwegian government organisation for aid and development (NORAD) was active before 1984. However, it has not been possible to get information about whether they had any activities in Bhopal after the leakage.

What we know is that many governments, including the Swedish, sent observers to Bhopal. The aim might have been catastrophe management and prevention as well as studying the effects of a toxic chemical.

It is said that many governments put pressure on the Government of India to accept the compensation sum offered by Union Carbide. This may have been one of the reasons for the very quick settlement in 1989.

Other companies like Mitsubishi and Bayer were said to have had information on MIC but maintained solidarity with UCC by not providing such information.

Around the times of the 12ᵗʰ and 13ᵗʰ anniversaries of the disaster, the European Economic Consortium (EEC) gave gifts of 300 and 200 tonnes of milk powder [28][16]. Distributed to 32,000 children in the severely affected communities, it would last for 1–2 months each year.

8.8.7 The role of non-governmental organisations

8.8.7.1 General aspects

The organisations that have been engaged in the Bhopal issue are numerous (appendix 4). Some of them existed before the Bhopal gas leak; others were started because of the consequences of the leak. Local, national and international networks have been formed.

The organisations are sometimes well-defined, sometimes loose networks. The local ones deal mainly with the rights of the survivors. National and international organisations and networks work within a broader field: in addition to the Bhopal issue they also focus on human rights, environmental issues, pesticides, education of adults, trade and transnational companies.

Several hundred or even thousands of articles and dozens of books on Bhopal have been published, demonstrations have taken place in different parts of the world, and several hundred thousand dollars have been collected in the western world. Activists have participated in the annual shareholders meetings of both UCC and Dow.

The international support is described by Jones [4]. The international networks were responsible for major investigations on Bhopal:

- The Highlander Research Document
- The reports in Business India
- The case study on double standards between UCC's plants in Bhopal and West Virginia
- The sessions of the Permanent Peoples' Tribunal [5][5]
- The sessions and reports of the International Medical Commission on Bhopal [32].

MacSheoin [9] points out that the international campaigns have had some success:

- Bhopal still remains an issue;
- The attempted out- of- court settlement in 1987 was scuttled;
- The reinvestigation of the criminal charges, which had been quashed in the February 1989 court settlement;
- Through raising the issue via shareholders, UC was forced to acknowledge continued water and soil contamination in Bhopal;
- While Dow initially refused to have any dealings with the victims of Bhopal, the demonstration in February 2001 forced it to enter negotiations.

Those events would not have been possible without the international networks that contributed economic support as well as experts.

8.8.7.2 The International Medical Commission on Bhopal

The story of the International Medical Commission on Bhopal (IMCB) in 1994 is an example of what an international professional group with restricted resources can do, and of the difficulties in handling such a group.

Fourteen doctors from 11 countries, with different specialities, plus a few other experts made up the IMCB. Over a three-week period, some limited studies were planned and conducted (see 8.2.8). No literature was distributed or recommended in advance. There was no time for enough education of the interviewers. As there was no ophthalmologist in the group, the participants were sent to a local doctor, but because of mistakes, the records were never analysed. For the immunological part, blood samples were collected but never analysed. It was suggested that a simulation test should be performed in the US, but the majority of the group was against this. No report was written on the contents of the cloud.

During this stay, press conferences and public meetings were held in Bhopal and New Delhi. In December 1996, when the final report was released, there were press conferences in New York, London, New Delhi and Bhopal. The results of the report have been presented at the Permanent People's Tribunal, in the medical press and at national and international medical conferences inside and outside India. Some of the material is found on the Internet.

There are some issues where the influence of the IMCB probably has been strong:
- The inclusion of neurotoxicity and post-traumatic stress disorder in the spectrum of gas-related diseases.
- The speeding up of the procedure of decisions on final compensation.
- The decision of the Bhopal Hospital Trust to supplement the hospital with outpatient clinics.

Because of differences of opinion about how the work should continue, IMCB has now been dissolved. However, individual members continue doing different kinds of work in Bhopal.

8.8.7.3 The International Campaign for Justice in Bhopal

The International Campaign for Justice in Bhopal (ICJB) is an umbrella organisation of all the groups who have joined forces to campaign for justice for the gas survivors of Bhopal [11]. The ICJB is spearheaded by survivors (the *Bhopal Gas Peedit Mahila*

Stationery Karmachari Sangh) and long-time supporters like the Bhopal Group for Information & Action, both Bhopal-based, and both plaintiffs in the ongoing Class Action suit in New York. International members include Greenpeace, Corpwatch, Essential Action and the Pesticide Action Network in the USA and UK.

Members of the ICJB jointly and separately campaign to secure justice for the survivors and those affected since by water supplies poisoned by toxic chemicals. They also campaign to put pressure on the governments of India and of the state of Madhya Pradesh to bring those responsible to trial, to ensure the clean-up of the contaminated factory and to distribute the funds meant for the survivors to the survivors. Mobilisation and coordination is made via the Internet.

In 2000, demonstrations were held in Mumbai and Delhi on the occasion of the visit to India of the then US President, Bill Clinton. The protest of the Bhopal survivors was represented both at the demonstrations against the World Bank and the IMF in Washington DC and Prague [26].

During 2002, a demonstration and hunger-strike were organised outside Dow Chemicals in Mumbai, with thousands of Bhopalis travelling there by train. By the time the hunger-strike ended, 1,500 people from 10 countries had joined the victims. Survivors' representatives travelled to Dow Chemical's offices around the world, and women gave brooms to the presidents – a great insult, and demanded a clean up. Sympathetic fasts were undertaken around the world.

At the 19th anniversary, 65 events (protests against Dow and Government of India) in 16 different countries were coordinated by ICJB.

In 2003, 18 members of the US Congress sent a letter to Dow Chairman, William Stavropoulos, demanding that his company assume liability for the wrongdoings of Union Carbide (its 100 percent subsidiary) in Bhopal [33, 34]. This is a result of the work from ICJB.

8.8.8 The role of trade unions

At the UC plant in Bhopal, the workers were organised into two competing trade unions. Management tried to use the rivalry to its advantage in contract negotiations [31]. The boards had to work hard to get the unions recognised [2]. Not until 1984 was the Union Carbide Karmachari Sangh recognised.

The workers' unions reacted as early as 1976, because of the pollution within the plant. Letters were sent to the managers of the plant and the factory inspector as well as to the Ministry of Labour of Madhya Pradesh [3]. They never received any answers.

After the dismissal of the two union leaders in 1982 (see 6.3.5), the union changed its focus, from the potential danger to all workers and the surrounding neighbourhood because of hazardous design, to the need to protect individual workers.

After the leak in 1982, the unions printed posters with warning texts that were distributed throughout the community.

It is not known whether the trade unions knew about UCC's plans to close down the plant. According to Indian law, the Government of India must sanction closing a factory and the dismissal of the employees. This sanction is only theoretical, because of the influence of the trade unions. The Indian National Trade Union Congress (INTUC) and other trade unions are very categorical about not changing this law.

The national trade unions did not take any active part in the protests after the leakage [4].

The alliance between gas workers and gas victims was more illusory than real. The trade union leaders were not arrested in the June 1985 midnight scoop (see 8.8.2). It was alleged that it was due to a deal with the police, that the unions would not take part in the demonstrations.

The Union Research Group in Mumbai formed the Trade Union Relief Fund, to support the workers' struggle for alternative food production at the UCC plant. In 1985, 400 people stormed into the plant to begin a sit-in to protest over job losses. The occupation of the plant did not end until December 1985, when UCIL made a large cash settlement with the workers [4]. However, the campaign was criticised on various grounds.

When the Bhopal Gas Affected Women's Stationery Workers' Union marched to Delhi in 1988 to fight for their rights, they got no support from the trade unions in Delhi [35].

Observers from two international trade unions came to Bhopal after the leakage. This resulted in the report "The Trade Union Report on Bhopal" [31], that was written "in response to our Indian affiliates".

Today, trade unions from across the world express solidarity with the cause of Bhopal gas victims. A joint appeal was issued by trade union representatives from over 25 countries in 2003 *(Hindustan Times, Bhopal lives, Dec 19, 2003)*.

8.8.9 The role of activists

Since 1984, individual activists have played a great role, as described by Jones [4]. The activists arriving in Bhopal soon after the accident often belonged to the left-wing opposition. This led to conflicts with other activists or NGOs, as well as with the police. Many left Bhopal within a few months, believing only interim relief was needed, and most had left Bhopal within a few years.

The first ever reports on the disaster were published by two activist organisations, Eklavya and the Delhi Science Forum.

The importance of women in the struggles after the killings cannot be overestimated. Women still dominate several of the national and local organisations.

Some activists in Bhopal did not want the questions of HCN and antidote treatment to grow too large. They maintained that it might reduce the question of the toxicity of MIC.

The conflicts with the police/MP government are described by Jones [4]. The activist groups realised early that the problems in Bhopal were too great for non-governmental organisations to cope with. Thus, rehabilitation activities were constantly coupled with attempts to organise the gas victims to demand their basic rights from the government. This led to violent repression from the police and the government.

There is still a group of international activists engaged in different issues: support to the survivors, the criminal case against Union Carbide, information, and prevention of "New Bhopals". Some of them visit Bhopal regularly and meet local people, an assurance to the locals that they are not forgotten.

At the moment, there is only one activist in Bhopal. On the other hand, he is a very important link to national and international networks. The Internet has become a very important way of information and communication.

8.8.10 Comments

8.8.10.1 Government ethics

The affected population was the poorest of the poor and a great proportion was Muslim. It is obvious that little interest in the welfare of the affected population has been shown by the authorities. It is likely that the attitude of the elite and the rulers towards the poorest is similar in countries with large material gulfs between different groups. In a country like India, the caste system and religious beliefs like reincarnation or Muslims' lower status might add to the neglect.

The class issue is relevant on a global as well as national level. The peoples of the poor countries are more exposed to environmental hazards than those of the rich countries. In each country, the poor parts of the population are more exposed that the rich sections.

Cassels [1] discusses the relations between transnational corporations and governments, especially in poor countries. He describes the complexity, and how dependent governments are on the corporations, that environmental and safety concerns will inevitably be balanced against the danger of the flight of capital. "What limited power the state does have to respond to the demands of workers and social activists is just that – limited."

8.8.10.2 Corporation ethics

OECD's guidelines for multinational corporations have existed since 1976 and are signed by 35 governments [36]. They were revised in 2000 to include environment and corruption, in addition to human rights, the rights of the employees, consumers interest, etc. The rules are voluntary, and there is no possibility of punishment.

Morehouse [8] concludes that UCC's behaviour was immoral. He also cites Lepkowski's words on the missed opportunity of the company to "display legal and moral innovation: a disaster one company decided not to turn its back on".

After the leakage, it seems obvious that UCC did their best to hide the truth. They firmly denied that MIC could have any long-term health effects. To support treatment with sodium thiosulphate would have confirmed that other compounds with serious and well-known effects on health had been released.

To keep the Bhopal plant in a sound and safe shape would not have involved a major investment for a large corporation like UCC. Paying their debts to the survivors would likewise not have been very costly for the corporation. It is clear that they regarded shares and money as more valuable than human beings.

Although UCIL has admitted that the sabotage theory was false [8], and the management's shortcomings are well known to the public, UCC still maintains that it was a disgruntled worker who caused the leakage. Since 1984, the consciousness ofn the companies' responsibility for environment and work environment has grown. How would the stock exchange react if UCC changed its message?

Dow Chemicals disputes all responsibilities for the victims and today's situation. This should be compared to ABB, the Swedish multinational that took over the American company Combustion Engineering around 1991. This company had until the 70s used asbestos as insulation material. ABB took over the compensation claims for those who fell ill. In 2001, it became possible also for persons who had not yet any symptoms to claim compensation. ABB has offered to pay more than 1 milliard dollars in compensation. (*Dagens Nyheter, Jan 18, 2003.*)

8.8.10.3 Medical ethics

The first known ethical rule for doctors is the Hippocratic Oath, which regulates the relationship between the physician and his patient. Also the Declaration of Medical Ethics of the World Medical Association (WMA) regulates the relations between the physician and his patients as well as his colleagues. The Helsinki declaration is on biomedical research ethics.

The code of medical ethics of the Medical Council of India, published at Medicon –96, is quite extensive. Here are some relevant points:

- The prime object of the medical profession is to render service to humanity; reward or financial gain is a subordinate consideration.
- The principle of the medical profession is to render service to humanity with full respect for the dignity of man.
- A physician should expose, without fear or favour, incompetent or corrupt, dishonest or unethical conduct on the part of members of the profession.
- Physicians, as good citizens possessed of special training, should give advice concerning the health of the community wherein they dwell. They should bear their part in enforcing the laws of the community and in sustaining the institutions that advance the interests of humanity. They should co-operate especially with the proper authorities in the administration of sanitary laws and regulations.

These codes are strong enough and would have great importance if they were followed.

Jones [4] has suggested that the fact that most Indian scientists are employed by the government and the public sector means they are wide open to state manipulation. On the other hand, in a country like Sweden, state manipulation is regarded as non-existent, and scientists employed by the public sector are assumed more neutral than those who are economically dependent on donors.

Many physicians in low-income countries are paid a very low salary by the state, which makes them susceptible to other sources of income.

There are some actions that might assist physicians to act in a more ethical way.

- The important parts of the Indian code of medical ethics could be an additional WMA declaration on the physician's duties to the public.
- One duty for WMA would be to support national medical organisations in creating and implementing this and other ethical codes.

- Another task for WMA would be to convince the WHO, the World Bank and the governments of the value of health professionals being paid decent salaries.

8.8.10.4 *The influence of non-governmental organisations and activists*

The local trade unions were the first to react to the hazards at the Bhopal plant. We do not know whether they were aware of the plans for closing down the plant. It is likely that there would have been forceful protests against the risks of unemployment, although from an environmental perspective, closing down the plant would have been absolutely the best action.

After the leakage, several thousand people in total must have been involved in work for the victims of Bhopal. Many of these are experts in different fields, and the total knowledge of the NGOs must be wide and deep. Many of the NGOs have contacts with the WHO and other organisations within the United Nations.

The commitment of all these NGOs has contributed to

- The Bhopal Gas Leak not being forgotten, either within or outside India;
- The speeding up of paying out interim relief and final compensation to the Bhopal victims;
- More regulations being created concerning hazardous industries, inside and outside India;
- The people of India becoming more conscious about the necessity of involving themselves in issues concerning environment and health.
- People outside India appearing to be more conscious about the Bhopal catastrophe today than in 1994.

All the same, it is remarkable that not more has been gained.

The survivors have been active and have organised themselves into various organisations. Unfortunately, the organisations seem to oppose each other more often than they co-operate. One of the activists from outside Bhopal expressed it this way: "The tragedy of the Bhopal Gas Leak includes the fights between the different survivors' organisations."

In English, two definitions are used: "victims" and "survivors". In my opinion, a "victim" is someone without power, who asks for help and looks for someone to blame. A "survivor" is someone who has the strength to manage a very difficult situation, and who now empowers him/herself. It seems as though the *Bhopal Gas Peedit Mahila Udyog Sangathan* tends to take on the role of "victim", while the *Bhopal Gas Peedit Mahila-Stationery Karmachari Sangh* acts more like "survivors".

For the self-identification of the people, it is probably not unimportant which word is used.

In contrast, compared to the survivors, the activists are often well educated and have large networks nationally and internationally. Without the activists, the survivors would not have gained as much as they have.

The risk with the involvement of activists is that all expectations are laid on their shoulders. When these are not fulfilled, or when the activists leave to work somewhere else, the survivors might feel abandoned and forgotten.

The wide-spread misunderstandings (see 8.6.1) that were partly supported by activists and NGOs might have created unnecessary fears and unrealistic expectations.

Less than one year after the Bhopal leak, two serious leaks took place at the plant in West Virginia. Many of the employees had believed UCC when they claimed that accidents like the one in Bhopal could not occur in America. This led to groups and organisations becoming aroused to work for workers' and environmental safety, and to demand for the right to know for employees and residents.

8.8.11 References

1. Cassels, J., *The Uncertain Promise of Law: Lessons from Bhopal.* 1993, Toronto: University of Toronto Press Inc.
2. *Other workers speak out: testimonies from Union Carbide Bhopal Plant personnel*, in *Bhopal et al. The Inside Story*, T.R. Chouhan, Editor. 1994, The Apex Press: New York.
3. Chouhan, T.R., *Bhopal: The Inside Story. Carbide workers speak out on the world's worst industrial disaster.* 1994, New York: The Apex Press.
4. Jones, T., *Corporate Killing. Bhopals Will Happen.* 1988, London: Free Association Books.

5. *Asia '92. Permanent Peoples' Tribunal. Findings and Judgements.* in *Third session on industrial and environmental hazards and human rights.* 1992, Oct 19–24. Bhopal-Bombay.
6. Lapierre, D. and J. Moro, *It Was Five Past Midnight in Bhopal.* 1st Indian ed. 2001, New Delhi: Full Circle Publishing. 376.
7. *Unraveling the Tragedy at Bhopal.* 1989, Union Carbide Corporation: USA.
8. Morehouse, W., *The Ethics of Industrial Disaster in a Transnational World: The elusive quest for justice and accountability in Bhopal.*
9. Mac Sheoin, T., *Report on Union Carbide Corporation.* 2002, The Other Media: New Delhi.
10. *Environment to the Ground and Bottom - experiences from the Halland ridge. Final Report from the Tunnel commission.* 1998, The Ministry of Environment: Stockholm. SOU 1998:137 [In Swedish].
11. *International Campaign for Justice in Bhopal ICJB.* 23 November 2002, www.bhopal.net.
12. Morehouse, W., *Unfinished business. Bhopal ten years after.* The Ecologist, 1994. **24(5)**.
13. Bidwai, P., *The poisoned city - diary from Bhopal,* in *Bhopal: Industrial genocide?* 1985, Arena Press: Hong Kong.
14. Chauhan, P.S., *Bhopal Tragedy. Socio-legal Implications.* 1996, Rawat Publications: Jaipur.
15. Morehouse, W. and A. Subramaniam, *The Bhopal Tragedy. What really happened and what it means for American workers and communities at risk.* 1986, New York: The Council on International and Public Affairs.
16. *Brief Descripiton of Work Done for Bhopal Gas Tragedy Relief,* I. Eckerman, Editor. 1995, Jan 9, Chadha, C.S. Principal Secretary: Bhopal.
17. *Bhopal Gas Tragedy Relief and Rehabilitation Department* http://www.mp.nic.in/bgtrrdmp/setup.htm. 2003, Government of Madhya Pradesh.
18. *Unsettling. What the Centre and the state government did to rehabilitate victims.* Down To Earth, 2003 (Dec 15): p. 33–34.
19. Shrivastava, S.B., *Letter on Conference on Bhopal Victims,* S. Acquilla. Jan 3, 2004, Bhopal Gas Tragedy Relief and Rehabilitation Department.
20. Sarangi, S., *Corporate Violence in Bhopal. International Conference on Preventing violence, Caring for survivors, Role of health profession and Service in violence.* 1998, Centre for Enquiry into Health and Allied: Mumbai.
21. *The Bhopal Gas Tragedy 1984– ? A report from the Sambhavna Trust, Bhopal, India.* 1998, Bhopal People's Health and Documentation Clinic: Bhopal.

22. *Charge Sheet*, Central Bureau of Investigations, DY. Supdt. of Police, CBI:ACU(I): New Delhi.

23. Jayaprakash, N.d., *Bhopal's Toxic Legacy. Tale of Gross Injustice*. 2003, Delhi Science Forum: New Delhi. p. 14.

24. *The Truth of Bhopal. Bhopal Gas Tragedy, 15th Anniversary*. 1999, Bhopal Gas Pidit Mahila Udyog Sanghthan, Stationery workers Union, Gas Pidit Avam Nirashrit Pension Bhogi Sangharsh Morcha: Bhopal.

25. *Commitments for Sustainable Development. Trade Unions at the Commission for Sustainable Development. Special "Business and Industry" Segment*. 1998, ICFTU: Brussels.

26. *Union Carbide Disaster in Bhopal. Fact sheet. 16th Anniversary of the December 2–3, 1984*. 2000, Bhopal Gas Pidit Mahila Udyog Sangathan and Bhopal Group for Information and Action: Bhopal.

27. *Carbide could escape court over Bhopal*. European Chemical News, 1996. **66**: p. 24.

28. *13th Anniversary Fact Sheet on the Union Carbide Disaster in Bhopal*, in *Compensation Disbursement*. 1997, Bhopal Group for Information and Action: Bhopal.

29. Agarwal, A., J. Merrifield, and R. Tandon, *No Place to Run. Local Realities and Global Issues of the Bhopal disaster*. 1985, Highlander Research and Education Center: Tenessee, USA.

30. *Bhopal Lives! Anniversary notes*. 1994, Bhopal Gas Peedit Sangharsh Sahayog Samiti and Bhopal Group for Information and Action: Bhopal.

31. *The Trade Union Report on Bhopal*. 1985, ICFTU–ICEF: Geneva, Switzerland.

32. Bertell, R. and G. Tognoni, *International Medical Commission, Bhopal: A model for the future*. Nat Med Journ India, 1996. **1996**(9): p. 86–91.

33. *Greenpeace India* www.greenpeaceindia.org. 2004.

34. *Corporate Watch India* www.corpwatchindia.org. 2004.

35. Eckerman, I., *Long-term Health Effects of Exposure to the Bhopal Gases. Observations of the hazardous effects on the health of particularly vulnerable groups*. 1995, Nordic School of Public Health: Goteborg, Sweden.

36. Saltas, A., *Can UN regulate the transnational companies? Let the UN-organ UNCTAD check the companies' direct investments in developing countries*. 2002, Sodertorns Hogskola: Stockholm. p. 18.

ANALYSIS & CONCLUSIONS

9.1 Analysis of Causes and Consequences

9.1.1 Injury analysis methods

Many models have been developed for analysing the extent of injuries [1]. Usually they are used for events like traffic accidents and children's burns.

The most well-known model is the Haddon model, which has three components: the causal chain of events, the Haddon matrix, the Ten technological strategies and the Four E's. The conception pre-event, event and post-event phases is used. L. R. Berger pointed out the limitations of the Haddon matrix: prevention is not emphasised, the social environment is hidden, and it is too complicated (personal communication). He has suggested a new model for prevention. The Logical Framework Approach [2] is a tool for project planning and management.

Originally, the Haddon and Berger models were used in this report, to analyse the causes of the disaster and its consequences. As a complement, the Logical Framework Approach was tested. As this model seems more complete and useful for this complex situation, it is the only model described here. However, the Haddon structure pre-event, event and post-event phases, is used for the structure of the book.

9.1.2 The Logical Framework Approach

The Logical Framework Approach (LFA) [2] is an analytical tool for objectives oriented project planning and management.

The key words are *objectives oriented, target group oriented* and *participatory*.

The LFA consists of the following parts:

1 *Participation analysis.* A comprehensive picture of the interest groups, the individuals and the institutions involved is developed.
2 *Problem analysis.* On the basis of available information, the existing situation is analysed. The major problems are identified and the main causal relationships between these are visualised as a problem tree.
3 *Objectives analysis.* The problem tree is transformed into a tree of objectives (future solutions to the problems) and analysed.
4 *Alternatives analysis.* Possible alternative options are identified, the feasibility of these is assessed and **one** project strategy is agreed upon.
5 *Developing the LFA matrix (matrices).* The main project elements are derived from the objective tree and transferred into goals, purpose, outputs, activities and inputs. Assumptions describe conditions that must exist, but which are outside the control of the project management. Indicators provide a basis for monitoring and evaluation and should specify target groups, quantity, quality, time and locations.

9.1.2.1

Table 6   *Identification of parties involved*

Governmental organisations
- Government of India
- Government of Madhya Pradesh
- Government of USA
- International governments
- WHO, United Nations

Commercial community
- UCC management
- UCIL management
- International trade organisations

Trade unions

- Trade unions at UCIL
- Trade unions in India
- Trade unions in West Virginia
- ILO

Civic community

- Residents in Old Town
- Residents in New Town
- Workers and operators at plant
- Survivors' organisations
- NGOs inside India
- NGOs outside India

Medical community

- Doctors at hospitals in Bhopal
- Other medical staff
- ICMR
- Private doctors in Bhopal
- Medico Friend Circle
- Indian Medical Association
- IMCB
- Scientists internationally

9.1.2.2 Problem tree

See Figure 2.

9.1.2.3 Tree of objectives

See Figure 3.

9.1.2.4

Table 7 *Matrix according to LFA*

Goal	Indicators	Assumptions
• Reduce the risk of the inhabitants being injured by toxic gases from plant	• No personal injuries from toxic gases from plant	• Appropriate laws • Appropriate funds • Interest from management • Interest from government • Residents interested
Purpose	**Indicators**	**Assumptions**
• Reduced risk of leakage • Reduced risk of being injured during leakage • Appropriate treatment if leakage occurs	• Fewer narrow escapes and small incidents • Residents have knowledge • No people living without houses or with poorly built houses close to plant • Health staff have knowledge and equipment • Employees, police, fire corps have knowledge about first aid	• Interest from UCC • Support from Governments
Outputs	**Indicators**	**Assumptions**
• Managers, operators and workers have adequate knowledge	• Tests	• Change of attitude of management • Redirection of investments

Continues...

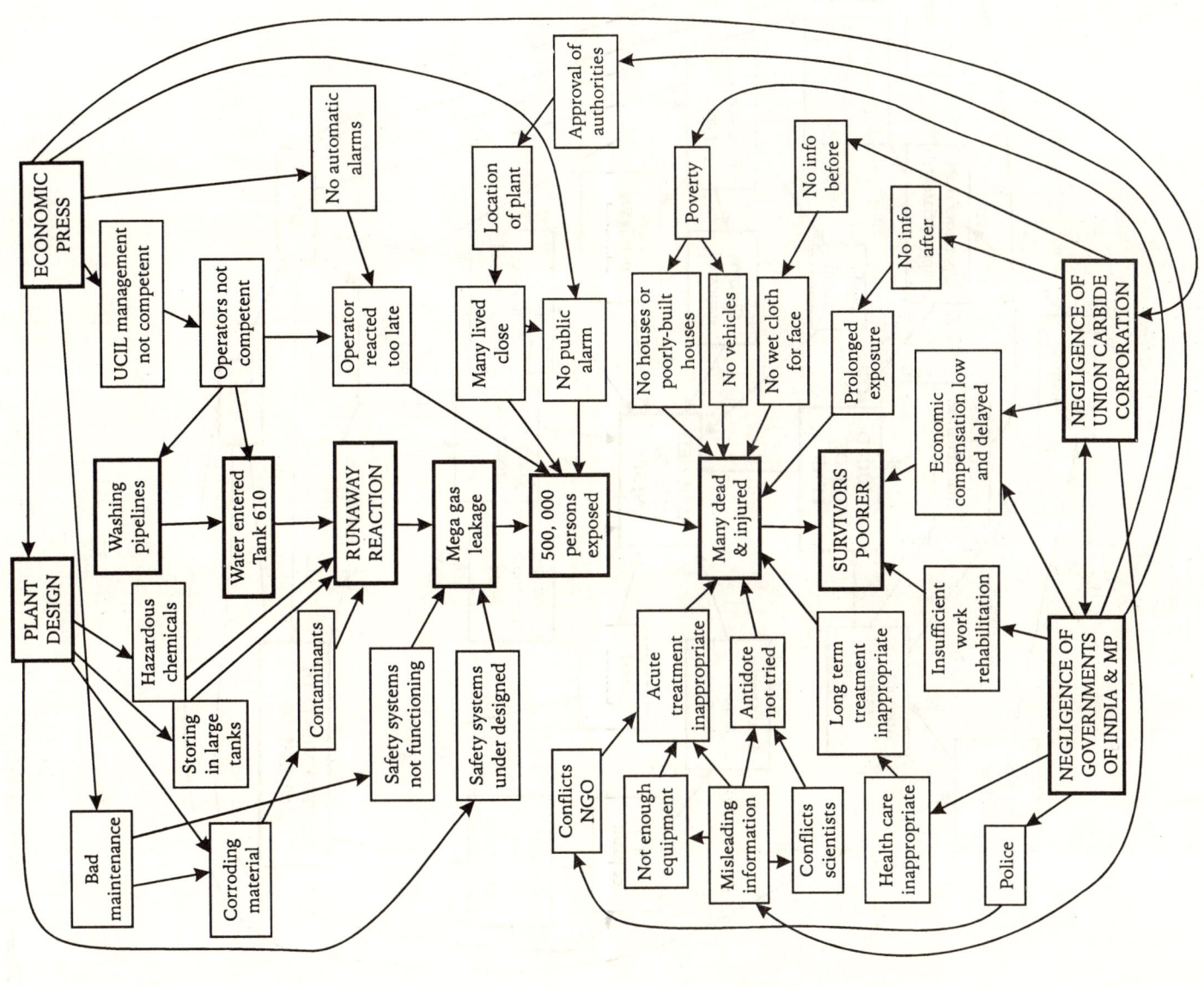

Figure 2   Problem tree

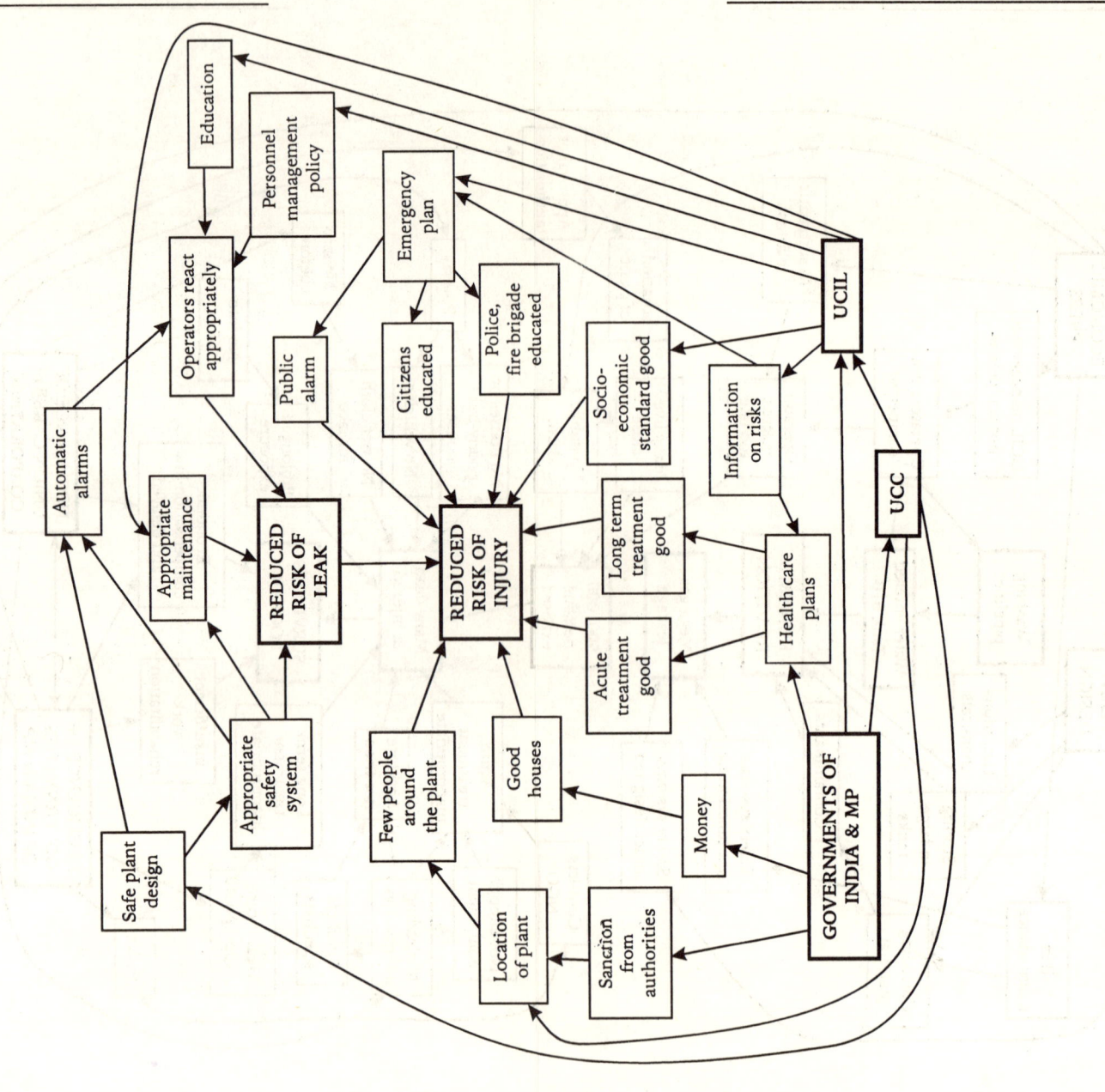

Figure 3 *Tree of objectives*

...Continued

• The plant in good shape at check-ups	• Protocol Investigation teams	
• All safety systems and rules always functional	• Protocol Investigation teams	
• Health staff have emergency plans	• Documented plans Practical exercises	
• Police, fire corps, authorities have emergency plans	• Documented plans Practical exercises	

Activities	Inputs	Assumptions
• Move factory?	• Money from UCC	• UCC interested
• Choose another method?	• Money from UCC	• Other methods exist and are cheap enough
• Change from storing in single large tanks to storing in several small tanks	• Money from UCC	
• Repairing all safety systems	• Skilled workers, money	
• Rules for safety controls	• Knowledge	
• Rules for maintenance	• Knowledge	
• Education programme for employees	• Time enough, knowledge	
• Education programme for inhabitants	• Teachers, money	
	• Teachers, money	
• Education programme for health staff, police, fire corps, authorities	• Teachers, money	

9.1.3 References

1. Berger, L.R. and D. Mohan, *Injury Control. A Global View*. 1996, Delhi: Oxford University Press.
2. *The Logical Framework Approach (LFA). Handbook for objective-oriented planning*. 1996, Norad: Oslo, Norway.

9.2 Conclusions

9.2.1 Analysis of methods

The Logical Framework Approach (LFA) seems a complete and useful model for a complex situation like the Bhopal gas leak. The problem and objectives trees look like a chain of events from where there are branches and roots. The matrix makes it possible to clarify what processes/changes from other instances are needed if the project is to succeed. As this is an analysis of an accident that has already happened, the matrix deals with both prevention and management. When planning a project, it may be clearer to create one matrix for prevention and another one or several others for management.

Despite my thorough knowledge of the Bhopal gas leak, developing this problem tree gave me some new insights on the connection between causes and effects. However, the tree looks more like a "problem net". When drawing the tree of objectives, I also acquired some new ideas on the measures necessary to prevent an accident or to mitigate its effects.

When visualising causes and consequences of this kind of accident, it is obvious that "chain" or "tree" is not the right word. "Net" is more appropriate.

9.2.2 Results of analyses

Analysis according to the LFA Problem Tree (see 9.1.2.2, Figure 2) demonstrates that to cause the mega-gas leak, it was not enough that water entered the tank. The most important factors were the plant design and the economic pressures.

The same analysis shows that the most important factor for the outcome of the leakage is the negligence of the Union Carbide Corporation and the Governments of India and Madhya Pradesh.

• The direct cause of the leakage is still unclear. However, the water washing theory seems most convincing.
• The direct cause of the leakage is less interesting, as the magnitude of the disaster was dependent on other factors.

- The parties responsible for the magnitude of the disaster are the two owners, Union Carbide Corporation and the Government of India, and to some extent, the Government of Madhya Pradesh.
- The leakage could have been prevented, even if the direct cause was sabotage.
- If the personnel management policy had been better, no "disgruntled worker" or "negligent employees" would have existed.
- The impact on health could have been reduced if the residents had been given information on how to behave in case of a leakage, and if they had been warned by the siren as soon as the leakage began.
- It is probable that closest to the factory, where the most serious health effects occurred, hydrogen cyanide was present in considerable concentrations. Therefore, early treatment with sodium thiosulphate would have mitigated the effects on the health of those living close to the factory.
- The effects on health caused by the leakage could have been mitigated if the medical, social, and economic rehabilitation had been adequate.
- The effects on health caused by the leakage could have been mitigated if the environmental rehabilitation had been adequate.

9.2.3 The roles of actors

9.2.3.1 General aspects

To reduce the influence on public health of chemical industries, there is a great need for action from many actors. The governments have a responsibility to protect their inhabitants from the negative effects of "development". As a result of globalisation, co-ordination between governments and national organisations is necessary.

Sweden is a country that has succeeded comparatively well concerning work environment, environment and human rights. The trade unions today are stronger than in most other countries. The non-governmental organisations are independent of the government, although they often get economic support from governmental institutions. The independent public service, such as governmental radio and TV channels, are considered as a guarantee to commercial and political interests. Corruption is low. The degree of education is high, and, what is more important, the income clefts are still rather small.

The following reflections are made from this – that it is possible to reach a balance between different interests, where the actors regard "the common utility" as something valuable. If it is possible in the Nordic countries, it should be possible also on a global level.

However, we must not forget that there has been a long struggle to reach this balance, and that it is very fragile. It needs continuous work from organisations and people to secure and develop this balance. Today, the struggle is global, in networks like the Social Forums that have been held all over the world during the last few years.

9.2.3.2 National and transnational companies

The boards of companies must realise that human beings have a value that is higher than the value of shares.

There is a need for ethical aspects to be incorporated in the visions of companies. It is becoming more common that companies report on environmental aspects in their annual reports. A chapter on ethical aspects is as necessary, including descriptions of the way in which each company reacted to ethical aspects. The guidelines of OECD can be included in the statutes of the corporations.

Companies must co-operate with the trade unions on work and environmental safety.

Companies operating in hazardous sectors must provide satisfactory guarantees of financial responsibility through minimum capitalisation and insurance requirements.

The companies must also enable and assist the recipient country to do a full-scale environmental impact assessment and risk analysis of the proposed installation, and to define the conditions under which the facility should operate.

9.2.3.3 Governments

It is the responsibility of each government to demand health consequence analyses before taking political decisions.

It is the responsibility of each government to establish strict health and safety standards for industries and to build up a control and punishment system.

This may come into conflict with international trends and "market" demands for reducing public expenses. The possibility of using taxes for covering the expenses may conflict with demands for support to companies. Government must co-operate in international networks, and support each other.

It is the responsibility of each government to provide their inhabitants with a health care system that they can trust.

The legislation system is a very important part of this. Qualified doctors and certified nurses guarantee a minimum standard, and it is possible to withdraw certification under certain circumstances.

Governments should regard non-governmental organisations as useful resources with a large knowledge base in certain fields.

Social-action and community groups can play a powerful role in the risk-management process. They are often best equipped to understand, articulate, and protect the interests of local communities. Too often, governments regard NGOs as obstacles.

Governments should regard trade unions as useful resources for the work with equality, unemployment and strikes.

The high grade of membership in trade unions in combination with central collective agreements leads to increased equality, less unemployment and fewer strikes [1].

9.2.3.4 The scientific and medical societies

The medical society has a moral duty to follow the ethical declarations of their organisations.

The scientific community has a moral duty to spread its knowledge to society, governments and others.

The international medical association has a moral duty to support colleagues in low-income countries to build up and maintain a high ethical level.

The power of the scientific and medical societies is very strong, but its potential is not used. It is necessary for scientists and medical persons to act independently of companies, political parties and others. This may be limited when companies instead of governments sponsor research, and when health care is privatised. In low-income countries, doctors are usually badly paid, and thus are susceptible to economic favours from companies. National medical organisations need support from international organisations.

9.2.3.5 *The international community*

International organisations, like the WHO and ILO, must provide guidelines on health and safety standards, as well as on disaster management, including epidemiology and economic compensation.

Cassels [2] discusses different ways of providing economic compensation to victims, irrespective of the cause. It is necessary to develop a system that is recognised as fair, and that is more expensive for the companies than preventing pollution.

World trade and economic organisations, including the World Bank, must realise that human beings have a value that is higher than the value of shares.

These organisations must listen much more to other actors. A system for health consequence analyses should be developed and used before any international agreement is made.

The world community must develop a model of public responsibility towards its less-developed members and, more generally, towards the global environment.

The Brundtland Report has identified as the most serious obstacle to achieving a safer and cleaner world as, "The relative neglect of economic and social justice within and amongst nations". It is suggested that UNCTAD (United Nations Conference on Trade And Development) could be strengthened and take this role [3].

The international society should regard the trade unions and non-governmental organisations as useful resources with a large knowledge base.

The NGOs have considerable expert knowledge in many fields. Information from NGOs and trade unions may often be very useful in supplementing information given by commercial companies and institutions.

9.2.3.6 *Non-governmental organisations*

Non-governmental organisations must co-operate with each other to gain strength.

NGOs are often small and have small resources. To be forceful, it is necessary that they unite their resources and not fight each other.

Non-governmental organisations must co-operate with governments and corporations, and lobby with the international society.

Non-governmental organisations must organise themselves so that the risk of corruption is minimised.

Many NGOs originate from movements on the left, and consider it against their policy to get involved with governmental or commercial institutions. But because of their small resources, most NGOs would gain more by working *with* the institutions instead of against them. As this may increase the risk of corruption, the NGOs must be organised so that the members have full insight and influence.

9.2.3.7 Trade unions

In general, what is good for work safety is also good for public and environmental safety. The trade unions must combine their forces to increase their influence on governments, companies and work places.

There is a demand from the companies that the power of trade unions should be reduced. In fact, in most parts of the world, the power of the trade unions needs to be strengthened. This requires co-operation between different unions instead of competition. It also means co-operation between scientists and non-governmental organisations and an understanding of the long-term interests.

However, it is not uncommon that trade unions protest vigorously when it is suggested that a polluting plant should be closed down.

9.2.3.8 The judiciary

The judicial community has a moral duty to follow the ethical declarations of its organisations.

The international judicial community has a moral duty to support colleagues in low-income countries to build up and maintain a high ethical level.

9.2.3.9 Survivors

Survivors' organisations must realise that they have more to gain if they co-operate than if they fight to maintain their own power.

Different organisations may be formed for different interest groups. However, the overall goals are probably the same. Uniting forces make the survivors stronger.

Survivors' organisations must be organised so that the risk of corruption is minimised.

The members must have full insight and a democratic influence in the organisation.

Survivors' organisations should use the experience, knowledge and support of other NGOs.

9.2.3.10 Activists

Activists must realise that they have more to win if they co-operate than if they fight for their own power.

Also, when activists come from different backgrounds, they must try to find the overall goals they have in common, and work together to reach those goals.

9.2.3.11 The public

The public must demand information on hazards to the environment.

The public must demand to be involved in the decision-making processes.

One problem in the democratic world is that the public does not use the opportunities they have of influencing their governments. The public must accept their responsibility for the environment.

The public must choose articles that are produced in a just and environmentally friendly way, and that are not hazardous to the environment.

Ultimately, it is the consumers who decide what articles are produced and how they are produced. If the public asks for food that is produced without pesticides, pesticide production will go down, and the risk of "new Bhopals" will decrease.

9.2.4 References

1. *Union and Collective Bargaining - Economic Effects in a Global Environment.* 2003, World Bank.
2. Cassels, J., *The Uncertain Promise of Law: Lessons from Bhopal.* 1993, Toronto: University of Toronto Press Inc.
3. Saltas, A., *Can UN regulate the transnational companies? Let the UN-organ UNCTAD check the companies' direct investments in developing countries.* 2002, Sodertorns Hogskola: Stockholm. p. 18.

9.3 Discussion

The Bhopal gas leak is not an exception. It is just the largest and most transparent chemical disaster that has occurred, until today. It clearly illustrates the threat to public health posed by the chemical industry:

- The regularly occurring leaks, explosions and other accidents, leading to a hazardous work environment and risk of exposure for the host population (appendix 3).
- The impossibility to foresee which chemical reactions and combinations of chemicals may arise during leaks and accidents (appendix 2).
- Direct damage to the environment during the production process, which creates hazards to human health.
- Production of substances, in this case pesticides, that are toxic to human beings when used, and are the cause of many deaths in large parts of the world.
- Production of substances that have long-term toxic effects on the environment, and which may lead to contaminated food and water as well as to decreased food production in the long run.

The Bhopal gas leak shows us the complexity of the chemical society we live in today. The fact is that we do not know which compounds we might be exposed to from chemical plants, and we do not know in what way and to what degree these compounds are harmful to us, in the short-term and in the long-term. If the "precautionary principle" was used all over, the number of chemicals would decrease dramatically.

It also illustrates the roles of transnational companies and how they succeed in influencing decisions at governmental and local levels. The same mechanisms are found everywhere, also in "rich" countries.

It is obvious that there are two antipodes concerning industrial hazards to public health: the industries on one side, and the trade unions and non-governmental organisations working for human rights and the environment on the other side.

The NGOs and trade unions usually fight for what is best for human beings. But we must realise that this is not the primary goal of a company. What is good for a company is not always good for the people.

For the people, and for public health, it is good that there are small income differences, strong working rights legislation, protection of water and ground, manpower-rich companies and the making of strong demands on the company concerning the work environment and the environment as a whole.

For the companies, it is good to have few employees, ease in exchanging labourers, an unsafe labour market which leads to the employed working hard and keeping silent, and low demands on the work environment and environment.

It is the task of the government to strike a balance between these two needs. For both the people and the companies, it is of course important that the company does well. The politicians should, nevertheless, put the interests of the people, both those now living and the coming generations, first.

The NGOs and trade unions could provide good support for the governments when trying to find out what is realistic and what is not concerning designs and safety measures. In Sweden, the taxes are quite high, but some of the taxes are paid to NGOs. For example, the Swedish Society for Nature Conservation gets economic support from the state, which gives it the opportunity to employ different kinds of experts as well as to inform the Swedes. The organisation is one of the instances for reviewing submissions on environmental questions for the government, and its influence is not negligible. The trade unions have had a very large impact on work safety and environment.

Taking care of our earth needs everyone's participation. Is there any hope that companies will participate? On the other hand, maybe they do not have any choice? To reduce the continuing degradation of our earth, there will be new rules of the game on the market. Will environmentally unfriendly companies survive in the long run?

10

REFERENCES AND STUDIED MATERIAL

10.1 Medical Scientific Material

10.1.1 Research on MIC

Goswami H K, Chandorkar M, Bhattacharya K, Vaidyanath G, Parmar D, Sengupta S, Patidar SL, Sengupta L K, Goswami R, Sharma P N. *Search for chromosomal variations among gas-exposed persons in Bhopal.* Hum-Genet, 1990 Jan;84(2):172–6

Chandra H, Rao G J, Saraf A K, Sharma V K, Jadhav R K, Sriramachari S. *GC-MS identification of MIC trimer: a constituent of tank residue in preserved autopsy blood of Bhopal gas victims.* Med Sci Law 1991;31:294–8.

Guest I, Varma DR. *Developmental toxicity of methylamines in mice.* J Toxicol Environ Health 1991;32(3):319–30.

Sriramachari S, Rao G J, Sharma V K, Jadhav R K, Saraf A K, Chandra H. *GC-NDP and GC-MS analysis of preserved tissue of Bhopal gas disaster: Evidence of methyl carbamylation in post-mortem blood.* Med Sci Law. 1991 Oct;31(4):289–93

Jeevaratnam K, Vaidyanathan C S. *Acute toxicity of methyl isocyanate in rabbit: in vitro and in vivo effects on rabbit erythrocyte membrane.* Arch Environ Contam Toxicol 1992;22:300–4.

Tamura N, Aoiki K, Lee M S. *Selective reactivities of isocyanates towards DNA bases and genotoxicity of methylcarbamoylation of DNA.* Mutat Res 1992;283:97–106.

Jeevaratnam K, Vidya S, Vaidyanathan C S. *In vitro and in vivo effect on methyl isocyanate on rat liver mitochondrial respiration.* Toxicol Appl Pharmacol 1992;117:172–9.

Dhara V R. *On the bioavailability of methyl isocyanate in the Bhopal gas leak* [editorial]. Arch Environ Health, 1992 Sept–Oct; 47(5):385–6.

Rennie J. Trojan horse. *Did a protective peptide exacerbate Bhopal injuries?* [News.] Sci Am, 1992 Mar;266(3):15–6.

Gupta G S, Bajpai R, Kaw J L, Dutta K K, Ray P K. *Modulation of biochemical and cytological profile of bronchoalveolar lavage constituents in rats following split-dose multiple inhalation exposure to methyl isocyanate.* Hum Exp Toxicol 1993;12:253–7.

Varma D R, Guest I. *The Bhopal accident and methyl isocyanate toxicity.* J Toxicol Environ Health, 1993 Dec;40(4):513–29.

Chandra H, Saraf A K, Jadhav R K, Rao G J, Sharma V K, Sriramachari S et al. *Isolation of an unknown compound, from both blood of Bhopal aerosol disaster victims and residue of tank E-610 of Union Carbide Limited. Chemical characterization of other structure.* Med Sci Law 1994;34:106–10.

Jeevaratnam K, Sriramachari S. *Comparative toxicity of methyl isocyanate and its hydrolytic derivates in rats I. Pulmonary histopathology in the acute phase.* Arch Toxicol 1994;69:39–44.

Sriramachari S, Jeevaratnam K. *Comparative toxicity of methyl isocyanate and its hydrolytic derivates in rats II. Pulmonary histopathology in the acute phase.* Arch Toxicol 1994;69:45–51.

Jeevaratnam K, Sriramachari S. *Acute histopathological changes induced by methyl isocyanate in lungs, liver, kidneys & spleen of rats.* Indian J Med Res 1994;99:231–5.

Jeevaratnam K, Vidya S. *In vitro and in vivo effects of methyl isocyanate on rat brain mitochondrial respiration.* Arch Environ Contam Toxicol 1994;27:272–5.

10.1.2 By different authors

Satyamala C et al. *Effect of Bhopal Gas Leak on Women's Reproductive Health.* Bombay: IBCS, 1985.

Satyamala C, Vohra N, Satish K. *Against All Odds. Continuing effects of the toxic gases on the health status of the surviving population in Bhopal.* New Delhi: CEC, 1989.

Distorted lives. Women's reproductive Health and Bhopal Disaster. Pune: Medico Friend Circle, 1990.

Mehta PS et al. *Bhopal tragedy's health effects. A review of methyl isocyanate toxicity.* JAMA 1990;264:2781–7.

Ghosh B B, Sengupta S, Roy A, Maity S, Ghosh S, Talukder G, Sharma A. *Cytogenetic studies in human populations exposed to gas leak at Bhopal, India.* Environ Health Perspect, 1990;86:323–6.

Bhandari, N R, Syal A V, Kambo I, Nair A, Beohar V, Saxena N C, Dabke A T, Agarwal S S, Saxena B N. *Pregnancy outcome in women exposed to toxic gases at Bhopal.* Indian J Med Res [B], Febr 1990, pp 28–33.

Kapoor R. *Fetal loss and contraceptive acceptance among the Bhopal gas victims.* Soc.Biol. 1991;38(3–4):242–8.

Kamat S R, Patel M H, Pradhan P V, Taskar S P, Vaidya P R, Kolhatkar V P, Gopalani J P, Chandarana J P, Dalal N, Naik M. *Sequential respiratory, psychologic, and immunologic studies in relation to methyl isocyanate exposure over two years with model development.* Environ Health Perspect, 1992 Jul;97–241–53.

Dhara V R. *Health Effects of the Bhopal Gas Leak: A Review.* Epidemiol/Prev 1992;14:22–31.

Sathyamala C. *Fertility and Gynaecological Disorders: Impact of Bhopal Gas Leak Disater.* London: Dep of Epidemiology and Population Sciences, London School of Hygiene and Tropical Medicine, 1993.

Narayan P S, Dwivedi M P, Sriramachari S, Agarwal P, Takle R. *Final Report of the Project on Pulmonary Function Test and Blood Gas Analysis.* Indian Council of Medical Research. Bhopal: Bhopal Gas Disaster Research Centre, Jawahar Lal Nehru Hospital, 1993.

Health Effects of the Bhopal Gas Leak Disaster. A review of the research projects of the Indian Council of Medical Research on the gas exposed population. New Delhi: Medico Friend Circle, 1994.

Dikshit R. *Lung, Oropharynx and Oral Cavity Cancer in Bhopal.* Dissertation. Tampere, Finland: University of Tampere, 1998.

Carlsten C. *An Eyewitness in Bhopal, 15 years after the disaster.* For BMJ, 2000. Not published.

Ranjan N, Sarangi S, Padmanabhan V T, Holleran S, Ramakrishnan R, Varma D R. *Methyl isocyanate exposure and growth patterns of adolescents in Bhopal.* JAMA, 2003; 290(14):1856–7.

10.1.3 By members of IMCB

Cullinan P, Acquilla S D, Dhara V R. *Long-term morbidity in survivors of the 1984 Bhopal gas leak.* Nat Med J India 1996;9:5–10.

Bertell R, Tognoni G. *International Medical Commission, Bhopal: A model for the future.* Nat Med Journ India 1996;9:86–91.

Bhatia R, Tognoni G. *Pharmaceutical use in the victims of the carbide gas exposure.* International Perspectives in Public Health (Buffalo, NY: Ministry of Concern for Public Health) 1996;11–12:14–22.

Eckerman I. *The health situation of women and children in Bhopal.* International Perspectives in Public Health (Buffalo, NY: Ministry of Concern for Public Health) 1996;11–12:29–36.

Eckerman I. *Long-term Health Effects of Exposure to the Bhopal Gases. Observations of the hazardous effects on the health of particularly vulnerable groups.* (Minor field study.) Goteborg: Nordic School of Public Health, 1995. Mfsrep.Int 1995:4.

Callender T. *Long-term neurotoxicity at Bhopal.* International Perspectives in Public Health (Buffalo, NY: Ministry of Concern for Public Health) 1996;11–12:36–41.

Heinzow B. *Results of the International Medical Commission on Bhopal (IMCB).* International Perspectives in Public Health (Buffalo, NY: Ministry of Concern for Public Health) 1996;11–12:4–8.

Jaskowski J, Zhengang W, Mohapatra S C, Bertell R. *Compensation for the Bhopal Disaster.* Int Perspectives in Public Health (Buffalo, NY: Ministry of Concern for Public Health) 1996;11–12:23–28.

Verweij M, Mohapatra S C, Bhatia R. *Health Infrastructure for the Bhopal Gas Victims.* Int Perspectives in Public Health (Buffalo, NY: Ministry of Concern for Public Health) 1996;11–12:8–13.

Cullinan P, Acquilla S, Dhara V R. *Respiratory morbidity 10 years after the Union Carbide gas leak at Bhopal: a cross sectional survey.* BMJ 1997;314:338–342.

Eckerman I, Gupta A, Durgavanshi S. *Effects of Yoga practices for respiratory disorders related to the Union Carbide Gas Disaster in 1984.* Buenos Aires: The XVI World Congress of Asthma, 1999.

Eckerman I. *Chemical Industry and Public Health. Bhopal as an example.* MPH 2001:24. Nordic School of Public Health, Gothenburg, Sweden: 2001.

Dhara V R, Dhara R, Acquilla S D, Cullinan P. *Personal Exposure and Long-term Health Effects in Survivors of the Union Carbide Disaster at Bhopal.* Environmental Health Perspectives, 2002; 110(51):487–500.

10.2 Material from ICMR

10.2.1 Manuals and the like

Health Effects on the Bhopal Gas Tragedy. New Delhi: Indian Council of Medical Research, 1986.

Working Manual 1 on the Health Problems of Bhopal Gas Victims. Assessment & Management. New Delhi: Indian Council of Medical Research, April 1986.

Bhopal Disaster. Manual of Mental Health Care for Medical Officers. Bangalore: ICMR centre for advanced research on community mental health, 1987.

The Health Problems of Bhopal Gas Victims. Assessment & Management. Working Manual 2. New Delhi: Indian Council of Medical Research and DST Centre for Visceral Mechanism, December 1989.

10.2.2 Annual reports

Indian Council of Medical Research. Annual Report 1990. Bhopal: Bhopal Gas Disaster Research Centre, Gandhi Medical College 1990.
Annual report 1991. Bhopal: Bhopal gas disaster research centre (ICMR)
Indian Council of Medical Research. Annual report 1992. Bhopal: Bhopal Gas Disaster Research Centre, Gandhi Medical College, 1992.
Indian Council of Medical Research. CoOnsolidated Report (summary). Bhopal: Bhopal Gas Disaster Research Centre, Gandhi Medical College, 1992.
Indian Council of Medical Research. Annual report 1993–94. Bhopal: Bhopal Gas Disaster Research Centre, Gandhi Medical College, 1994.

10.3 Material from Indian Official Authorities

10.3.1 Government of Madhya Pradesh

Government of Madhya Pradesh. *Census 1981.* Bhopal: The Census Department, 1982.
Government of Madhya Pradesh. *Census 1991.* Bhopal: The Census Department, 1991.
Medical Documentation for Claimants. Guidelines for Medical Officers. Bhopal: Bhopal gas tragedy relief & rehabilitation department.
Medical documentation for claimants. Guidelines for medical officers volume II.
Special Industrial Area Bhopal (Gas relief). Industrial Opportunities. Bhopal: MP Audyogik Kendra Vikas Nigam (Bhopal), Ltd, 1988.
Community Health Scheme. An Introduction. Bhopal: Bhopal Gas Tragedy Relief & Rehabilitation Department, Gov of MP, 1991.
Socio-economic Impact of Disbursement of Interim Relief to Gas-affected Families of Bhopal. 1991, Academy of Administration, Government of Madhya Pradesh: Bhopal. p. 154.
Brief descripiton of work done for Bhopal gas tragedy relief. Written information received from Mr C.S. Chadha, Principal Secretary, Bhopal Gas Tragedy Relief Department, Government of Madhya Pradesh, on 9th January, 1995.

10.3.2 Central Bureau of Investigation

Central Bureau of Investigations. Chargesheet. New Delhi: DY. Supdt. of Police, CBI:ACU(I).

10.3.3 Disaster Management Institute

Disaster Management Institute. Information brochure. Bhopal: Disaster Management Institute, 1994.
Refresher Course for Top Executives. Management of Chemical Accidents. Bhopal: Disaster Management Institute, 1994.
- Sengupta M. *Emergency Preparedness Plan for Chemical Hazards.*
- Raghupathy L. *Waste Minimisation: An Approach.*
- Mittal B.K. *Chemical Health Hazards & Their First-aid Measures.*
- Ramachandran CR. *Immediate Post-industrial Disaster Management*
Choukse RM. *Safety Requirements under Factories Act and Rules Pertaining to Chemical Accidents.*

10.3.4 Gandhi Medical College

Singh S. *Chemistry, Fate, Pharmacology & Effects of Methyl isocyanate. Only for Medical Profession.* Bhopal: Department of Pharmacology, Gandhi Medical College, 1985.
Report on family welfare/primary health care needs of slumdwellers of Bhopal town and formulation of proposal for strengthening of existing family welfare services and creation of new health facilities. Bhopal: Department of Preventive and Social Medicine, Gandhi Medical College, 1992.

10.3.5 Council for Scientific and Industrial Research

Varadarajan S et al. *Report on Scientific Studies on the Factors Related to Bhopal Toxic Gas Leakage.* New Delhi: Indian Council for Scientific and Industrial Research, 1985.

10.4 Material from Union Carbide

10.4.1 Manuals published before 1984

Methyl isocyanate. Union Carbide F-41443A – 7/76. New York: Union Carbide Corporation, 1976.
Carbon monoxide, Phosgene and Methyl isocyanate. Unit Safety Procedures Manual. Bhopal: Union Carbide India Limited, Agricultural Products Division, 1978.

Behl VK, Agarwal VN, Choudhary SP, Khanna S. *Operating Manual Part-II. Methyl Isocyanate Unit*. Bhopal: Union Carbide India Limited, Agricultural Products Division 1979.

10.4.2 Material published after 1984

Bhopal Methyl Isocyanate Incident. Investigation Team Report. Danbury, CT: Union Carbide Corporation, 1985.

Kalelkar AS, Little AD. *Investigation of Large-magnitude Incidents: Bhopal as a Case Study*. Presented at the Institution of Chemical Engineers conference on preventing major chemical accidents. London, 1988.

Presence of Toxic Ingredients In Soil/Water Samples Inside Plant Premises. USA: Union Carbide Corporation, 1989.

Unraveling the Tragedy at Bhopal. Videotape. USA: Union Carbide Corporation, 1989.

Browning JB. *Union Carbide: Disaster at Bhopal*. USA: Union Carbide Corporation, 1993.

Union Carbide Corporation. Bhopal Fact Sheet. USA: Union Carbide Corporation, 1995.

10.5 Material from Bhopal Memorial Hospital Trust

The Bhopal Memorial Hospital and Research Centre and Outreach Health Centres. Annual report 2001. Bhopal: Bhopal Memorial Hospital Trust, 2001.

10.6 Material Published by NGOs

10.6.1 Bhopal Group for Information and Action (BGIA)

Voices from Bhopal. Bhopal: Bhopal Group for Information and Action, 1990.

Compensation Disbursement. Problems and Possiblities. Bhopal: Bhopal Group for Information and Action, 1992.

- Lochan R. *Health Damage Due to Bhopal Gas Disaster. Review of Medical Research.*
- *13th Anniversary Fact Sheet on the Union Carbide Disaster in Bhopal.*

Long-term Epidemiological Studies on the Health Effects of Toxic Gas Exposure through Community Health Clinics. A Summary. Bhopal: Bhopal Group for Information and Action, 1994.

Clinical profile of the Bhopal Gas Victims. A Summary of ICMR reports. Bhopal: Bhopal Group for Information and Action, 1994.

Bhopal Lives! Anniversary Notes. Bhopal: Bhopal Gas Peedit Sangharsh Sahayog Samiti and Bhopal Group for Information and Action, 1994.

Statement of Bhopal Peedit Mahila Stationery Karmachari Sangh before the IMCB. Bhopal: Bhopal Group for Information and Action, 1994.

Narayan T. *Submission on the health status and health care of victims of the Bhopal gas disaster of 1984, to the International Medical Commission on Bhopal.* Bhopal: Medico Friend Circle and Bhopal Group for Information and Action, 1994.

Sarangi S. *Corporate Violence in Bhopal.* International Conference on Preventing violence, Caring for survivors, Role of health profession and Service in violence. Mumbai: Centre for Enquiry into Health and Allied, 1998.

Union Carbide Disaster in Bhopal. Fact Sheet. 16th Anniversay of the December 2–3, 1984. Bhopal: Bhopal Gas Pidit Mahila Udyog Sangathan and Bhopal Group for Information and Action, 2000.

10.6.2 Sambhavna Trust

Sarangi S. *Methyl isocyanate (MIC).* Bhopal: Bhopal People's Health and Documentation Clinic (Sambhavna Clinic).

Report of Survey for Assessment of Drug Distribution in Gas-affected Bhopal. Bhopal: Sambhavna Clinic.

Prevalence of diabetes mellitus, hypertension and under/over-weight among the people exposed to toxic gases from UCIL. Bhopal: Sambhavna Clinic.

Report from the Sambhavna Clinic, Bhopal I. Bhopal: Sambhavna Trust, 1997

The Bhopal Gas Tragedy 1984– ? A report from the Sambhavna Trust, Bhopal, India. Bhopal: Bhopal People's Health and Documentation Clinic, 1998.

Assessment of treatment offered at the Bhopal Hospital Trust's community clinic no.1 and analysis of Bhopal Hospital Trust prescription data. Bhopal: Sambhavna Clinic, 1998.

Gupta A, Durgvanshi S. *Yoga therapy in the care of chronic respiratory disorders among survivors of the December 1984 Union Carbide disaster in Bhopal.* Presentation at the 3rd International Conference on "Yoga Research & Traditions", Kaivalyadhjama, Lonavia, Jan 1–4, 1999.

Gupta A, Durgavanshi S, Eckerman I. *Effects of Yoga practices for respiratory disorders related to the Union Carbide Gas Disaster in 1984.* Buenos Aires: The XVI World Congress of Asthma, 1999.

Stop the Ongoing Medical Disaster. Bhopal Healthcare is Sick. Bhopal: Sambhavna Trust, 2000.

Carlsten C. *Critical Issues in Diagnosis and Treatment of Common Presentations in Bhopal India.* Bhopal: Sambhavna Trust, 2000.

Menstrual Problems Reported by Teenagers Registered at Sambhavna Clinic. Bhopal: Sambhavna Trust, 2000.

Welcome to Bhopal and the Sambhavna Clinic. In: Material for the visit of Dominique Lapierre and Javier Moro. Bhopal: Sambhavna Trust, 2001.

Ranjan N, Sarangi S, Padmanabhan VT, Holleran S, Ramakrishnan R, Varma DR. *Methyl Isocyanate Exposure and Growth Patterns of Adolescents in Bhopal.* JAMA, 2003;290(14):1856–7.

Press Statement, Jan 30, 2003. Bhopal: Sambhavna Trust, 2003.

10.6.3 Other NGOs

Agarwal A, Merrifield J, Tandon R. *No Place to Run. Local Realities and Global Issues of the Bhopal Disaster.* Tenessee, USA: Highlander Research and Education Center, 1985.

Information to Gas-affected Women in Bhopal. (In Hindi.) Bhopal: Eklavya, 1986

No more Bhopals. The Survivors Speak Out. London: Bhopal Action Group/ TICL, 1989.

Union Carbide in Bhopal, India. The Lingering Legacy. Analyses of Carbide-related Toxins at the Former UCIL site. Boston: National Toxics Campaign Fund, 1990.

Encroachment on Civil Rights. Report of an investigation into "Anti-encroachment drive". Bhopal: People's Union for Civil Liberties, MP, 1991.

Contempt of People. Ramifications of the Bhopal Gas Leak Disaster Case. New Delhi: Delhi Science Forum, 1992.

Asia '92. Permanent People's Tribunal. Findings and judgements. Third session on industrial and environmental hazards and human rights: 19–24 October 1992, Bhopal-Bombay (India).

Bhopal: The second tragedy. Birmingham: Central Broadcasting, 1995.

Charter on Industrial Hazards and Human Rights. Permanent People's Tribunal. London: PesticideTrust, 1996.

The Truth of Bhopal. Bhopal Gas Tragedy, 15th anniversary. Bhopal: Bhopal Gas Pidit Mahila Udyog Sanghthan, Stationery Workers Union, Gas Pidit Avam Nirashrit Pension Bhogi Sangharsh Morcha, 1999.

Proposal to set up office-stationery production. Bhopal, 2000.

10.7 Articles from Non-scientific papers

Kumar, K., *Princes and pesticides*. South Asia - Political and cultural magazine 1985; 3:7–8. [In Swedish]

Gunnarsson, B., *The fight for justice*. South Asia - Political and cultural magazine 1985;3:3–6. [In Swedish]

Malmberg, C., *The whole third world is poisoned*. South Asia–Political and cultural magazine 1985;3:9–11. [In Swedish]

Subramaniam A. *Bhopal - The dangers of diagnostic delay*. Business India 1985; August;12–25.

Worthy W. *Methyl isocyanate: The Chemistry of a Hazard*. Chemical and Engineering News, Chicago 1985;Febr 11, 27–33

Mohan D, Sadgopal A, Verma V. *Disaster management. Lessons from Bhopal*. Business India 1987;Nov30–Dec13:76–85.

Mohan D. *Disaster by plan*. Lokayan Bulletin 1988;6 1/2:141–6.

Sathyamala C. *The medical profession and the Bhopal tragedy*. Lokayan Bulletin 1988; 6:1/2.

Jaising I, Sathyamala C. *Legal rights and wrongs. Internationalising Bhopal*. Development Dialogue 1992:1–2.

Morehouse W. *Unfinished Business. Bhopal ten years after*. The Ecologist 1994;24(5)

Lepkowski W. *Bhopal - ten years later*. Washington: C&EN 1994;Dec 19.

Carbide could escape court over Bhopal. European Chemical News, 1996;66:24.

Mehta S. *Bhopal lives*. Village Voice, 1996;Dec 3.

Mathew J. Fighting for a cause. Chronicle, 1999;May 9

Gas victims brave dry taps in filthy colony. Hindustan Times, 2001;Jan 22.

Serving the symptom. The government still does not know what still afflicts people in Bhopal. Down To Earth, 2003;Dec 15:29–32.

Unsettling. What the Centre and the state government did to rehabilitate victims. Down To Earth, 2003;Dec 15: 33–34.

Foul Debris. The UCIL plant is still a health hazard. Down To Earth, 2003;Dec 15.

Hinterland. A Special Issue on the Bhopal Gas Tragedy. New Delhi: Department of English, Hindu College 2003.

- Mukherjee S, *The Lessons of Collective Action in Bhopal. An Interview with Sathyu.*.

Gas victims brave dry taps in filthy colony. Hindustan Times Bhopal, 2001;Jan 22.

10.8 Reports

Morehouse W. *The Ethics of Industrial Disaster in a Transnational World: The elusive quest for justice and accountability in Bhopal.*
Description on possible compounds of the gas cloud. Title unreadable. No author, no date.
Morehouse W, Subramaniam A. *The Bhopal Tragedy.* New York: Council on International and Public Affairs, 1985.
The Trade Union Report on Bhopal. Geneva, Switzerland: ICFTU–ICEF, 1985.
Karlsson E, Karlsson N, Lindberg G, Lindgren B, Winter S. *The Bhopal Catastrophe: Consequences of a Liquified Gas Discharge.* Umea, Sweden: Forsvarets Forskningsanstalt (National Defence Research Institute), 1985. C40212–C1(C2,C3,H1).
Morehouse W, Subramaniam A. *The Bhopal Tragedy. What really happened and what it means for American workers and communities at risk.* New York: The Council on International and Public Affairs, 1986.
Shelley H. et al. *The Health Effects of Methyl isocyanate, Cyanide and Monomethylamine exposure.*
Kulling P, Lorin H. *The Toxic Gas Disaster in Bhopal December 2–3, 1984.* Stockholm: Forsvarets Forskningsanstalt (National Defence Research Institute), 1987. KAMEDO 0281–2223:53. [In Swedish]
The Bhopal Legacy. Toxic contaminants at the former Union Carbide factory site, Bhopal, India: 15 years after the Bhopal accident. Exeter: University of Exeter, 1999.
Meenakshi, N. *Industrial Disasters Working Towards Oblivion.* In: India disasters report. Towards a policy initiative. Ed Parasuraman S, Unnikrishnan PV. New Delhi: Oxford University Press, 2000
A Report on Mercury Contamination of Groundwater Near the Union Carbide Factory at Bhopal. Dehra Doon: Peoples' Science Institute, 2001.
Mac Sheoin T. *Surviving Bhopal: 16 Years on a Fact-finding Mission. Report on Union Carbide.* Dublin, 2001.
Mac Sheoin T. *Report on Union Carbide Corporation.* New Delhi: The Other Media, 2002.
Murthy R S. *Mental Health Impact of Bhopal Gas Disaster.* New Delhi: The Other Media, 2002.
Srishti. *Surviving Bhopal. Toxic present – Toxic Future. A Report on Human and Environmental Chemical Contamination around the Bhopal disaster site.* New Delhi: The Other Media, 2002.
Saltas, A., *Can UN Regulate the Transnational Companies? Let the UN-organ UNCTAD check the companies' direct investments in developing countries.* Stockholm: Sodertorns Hogskola, 2002 [InSwedish]

Jayaprakash, N.D. *Bhopal's Toxic Legacy. Tale of Gross Injustice*. New Delhi: Delhi Science Forum, 2003.

Union and Collective Bargaining - Economic Effects in a Global Environment. World Bank, 2003.

10.9 Books

Bhopal Gas Tragedy. New Delhi: Delhi Science Forum, 1985.

Weir D. *The Bhopal Syndrome. Pesticides, Environment and Health*. London: Earthscan publications Limited, 1988.
  • Alvares C: *A Walk through Bhopal*.

Cassels J. *The Uncertain Promise of Law: Lessons from Bhopal*. Toronto: University of Toronto Press Inc, 1993.

Pandey AK. *The Ophidian and the Orphans of Bhopal. Bhopal Gas Tragedy*. Bhopal: Rajdhani Law House, 1994.

Chauhan PS. *Bhopal Tragedy. Socio-legal Implications*. Jaipur: Rawat Publications, 1996.

Lapierre D, Moro J. *It Was Five Past Midnight in Bhopal*. New Delhi: Full Circle Publishing, 2001.

Chouhan TR et al. *Bhopal: The Inside Story. Carbide workers speak out on the world's worst industrial disaster*. New York: The Apex Press, 1994.
  • *Other Workers Speak Out: testimonies from Union Carbide Bhopal Plant personnel*.
  • Alvares C. *Bhopal Ten Years After*.
  • Jaising I. *Legal Let-down*.

Silent invaders. Pesticides, Livelihoods and Women's Health. Ed Jacobs M, Dinham B. London: Zed Books Ltd, 2003.
  • Sarangi S. *The Bhopal aftermath: generations of women affected*.

MacSheoin, T, *Asphyxiating Asia*. Mapusa, Goa: Other India Press, 2003.

10.10 Background Material

10.10.1 Toxicology

Patty FA ed. *Industrial Hygiene and Toxicology*. Second Revised Edition. Vol 2. New York: Interscience Publishers, John Whiley & Sons, 1963.

Occupational Health Guideline for Methyl isocyanate. Washington DC: US department of Health and Human Services, September 1978.

A Recommended Standard for Occupational Exposure to Di-isocyanates. Washington DC: US Department of Health, Education and Welfare, 1979.

Clayton GD, Clayton FE ed:s. *Patty's Industrial Hygiene and Toxicology.* Third revised edition. Vol 2A, 2B, 2C. New York: Wiley-Interscience Publication, John Wiley & sons, 1982.

Lorin HG, Kulling PEJ. *The Bhopal tragedy - what has Swedish disaster medicine learned from it?* J Emerg Med 1986;4:311–6.

Lundquist P. *Determination of Cyanide and Thiocyanate in Humans.* Linkoping, Sweden: Department of Clinical Chemistry, Linkoping University, 1992.

Karlson-Stiber C, Hojer J, Sjoholm A, Bluhm G, Salmonson H. *Toxic pneumonitis in ice-hockey players.* Lakartidningen 1996;93:3808–12. [In Swedish]

Theme: *How dangerous are the isocyanates?* Aktuell arbetslivs-forskning (Stockholm: Swedish Council for Work Life Research) 1997;6:3–15. [In Swedish.]

Scientific basis for threshold limit values. Ed Montelius, J. Work and Health, 2001:19. Stockholm: National Institute for Working Life, 2001. [In Swedish]

10.10.2 Pathology

Haeggstrom F, Eklund A, Elmberger G. *Distinguish beteween BO and BOOP! Two pulmonary diseases that often are mixed up.* Lakartidningen 1997;94:1287–91. [In Swedish]

10.10.3 Injuries

Berger LR, Mohan D. *Injury control. A global view.* Delhi: Oxford University Press, 1996.

The Logical Framework Approach (LFA). Handbook for objective-oriented planning. Oslo, Norway: Norad 1996.

Perrow C. *Accident in high-risk systems.*

10.10.4 Post-traumatic stress disorder, night work

Mitler MM, Carskadon MA, Czeisler CA, Dement WC, Dinges DF, Graeber RC. *Catastrophes, sleep and public policy: Consensus report.* Sleep, 1988;11:100–9.

Sondergaard, HP: *Traumatic Stress.* Lakartidningen 1993;90:796–800. [In Swedish]

Akerstedt T. *Increased risk for accidents at night work.* Lakartidningen, 1995;92:2103–4. [In Swedish]

Koscheyev VS, Leon GR, Gourine AV, Gourine VN. *The psychosocial aftermath of the Chernobyl disaster in an area of relatively low contamination.* Prehospital and Disaster Medicine, 1997;12:41–6.

Abdulbaghi A, Sundelin Wahlsten V. *Child trauma and vulnerability; new knowledge necessitates conceptual revision.* Lakartidningen 1998; 17:1955–62. [In Swedish]

Dotevall G. *Stress and Psychosomatic Disease.* Lund, Sweden: Studentlitteratur, 2001. [In Swedish]

10.10.5 Socio-economic conditions and health

Households and Health. In: World Development Report 1993. Investing in Health. New York: Published for the World Bank by Oxford University Press, 1993.

Poverty and Disability. A position paper. Stockholm: Swedish International Development Authority (SIDA), 1995

Globalisation – What is That? Stockholm: Forum Syd, 1997. [In Swedish]

10.10.6 Environmental risk management

Environment to the Ground and Bottom – Experiences from the Halland ridge. Final report from the Tunnel commission. Stockholm: The Ministry of Environment, 1998. SOU 1998:137. [In Swedish]

Lindgren L. *Sustainable Development. Children and Environment.* Stockholm: Save the Children, 1998. [In Swedish]

Tropp H. Patronage, Politics and Pollution. *Precarious NGO-State Relationships: Urban Environmental Issues in South India.* Linkoping, Sweden: The Tema Institute, Linkoping University, 1998.

Commitments for Sustainable Development. Trade Unions at the Commission for Sustainable Development. Special "Business and Industry" Segment. Brussels: ICFTU, 1998.

10.10.7 Human rights

The International Bill of Human Rights. New York: United Nations, 1993.

The Montreal Declaration on People's Right to Safety. 6th World Conference on Injury Prevention and Control, Montreal, Canada, 15th May 2002 http://www.iitd.ac.in/tripp/righttosafety/Montreal%20declaration%2015–05–02.html

10.11 Web Sites

Call for Dissolution of Union Carbide. Pesticide Action Network North America (PANNA), Mar 05, 1996

Taqi A. *Globalization of Economic Relations: Implications for Occupational Safety and Health: An International View.* International Labour Organization (ILO), Apr 23, 1996.

Goswani M. *Second Generation Victims of Bhopal Tragedy – New Medical Centre to Provide Succour.* Business Lin & The India Information Inc, Oct 13, 1996.

13th Anniversary of the Bhopal Disaster. Press Release from Bhopal Gas Peedith Sangarsh Sahyog Samiti. Corporate Watch, Dec 1997

The Indian Express, 1998, Oct 02.

Bhopal gas victims stage demonstration. The Hindu, Dec 05, 1998.

Bhopal: setting the record straight. A conversation with Edward A Munoz, former Managing Director of Union Carbide India, Limited. Corporate Watch, 1998.

Sharma ND: *New gas victims hospital opens, the process plods on.* The Indian Express, 1998.

Noronha F. *India to Build Memorial to Bhopal Tragedy.* Hoosier Environmental Council (HEC), July 17, 1998.

The Machine Stops: Runaway Reaction. 2000/01/05 http://home.earthlink.net/~wroush/disasters/bhopal3.html.

International Campaign for Justice in Bhopal ICJB www.bhopal.net.

Bhopal Medical Appeal www.bhopal.org

Greenpeace International www.greenpeace.org/campaigns

Greenpeace India www.greenpeaceindia.org

Bhopal www.bhopal.com Copyright Union Carbide Corporation

Corporate Watch India www.corpwatchindia.org

Government of Madhya Pradesh. Bhopal Gas Tragedy Relief and Rehabilitation Department http://www.mp.nic.in/bgtrrdmpsetup.htm

10.12 Personal Communication

Acquilla, S. and R. Dhara, *Interview with Dr. Banerjee and Moina Sharma*, *Center for Rehabilitation Studies*. 2004, Personal communication.
Shrivastava, S B. *Letter to Acquilla, S on Conference on Bhopal victims*. Bhopal Gas Tragedy Relief and Rehabilitation Department: Jan 3, 2004.

11

ACKNOWLEDGMENTS

First of all, I want to thank my supporters in India:
- Satinath Sarangi (Sathyu), the "fire soul" who introduced me to the story and provided me with contacts and material.
- The Siddiqui family, who adopted me, and made it important for me to keep on visiting Bhopal.
- The staff at Sambhavna, who always made me feel welcome.
- Rashida Bee and Kampa Devi Shukla, the presidents, and the members of *Bhopal Gas Peedit Mahila-Stationery Karmachari Sangh*, who willingly let me interview them in my second study.
- Mr Abdul Jabbar Khan, the president of *Bhopal Gas Peedit Mahila Udyog Sangathan.*
- Mira Sadgopal, Mira Shiva, Sathyamala and the other doctors of Medico Friend Circle, who helped and supported me in my work for IMCB, 1994.
- Chouhan, author of *"Bhopal. The inside story"* that is full of useful information.
- Deeno, former member of *Bhopal Group for Information and Action.*
- Late Dr Dwivedi, who gave me information on ICMR.
- Deenadaylan and his wife, who invited me to their home in New Delhi.
- The staff at the Other Media, who never got tired of arranging hotels and train tickets for me.
- Jayaprakash from Delhi Science Forum, who sometimes managed to see me before I left for Sweden.
- Dinesh Mohan at IIT in New Delhi, who taught me injury prevention.

Next, I would like to thank my encouraging supporters in Sweden:

- Vinod Diwan at NHV, originally from Bhopal, who has been my supervisor
- Ragnar Andersson, earlier at Community Medicin, County of Stockholm, who also has been my supervisor
- Frants Staugaard, earlier at NHV, who encouraged me to even start this kind of work
- Hans Rosling, earlier at ICH in Uppsala, who taught me health care in low income countries
- Bertil Egero, my brother who understood what I was doing
- Madi Gray, who has done all the language checking

APPENDICES

12.1 Appendix 1: Map of Distribution of the Gases

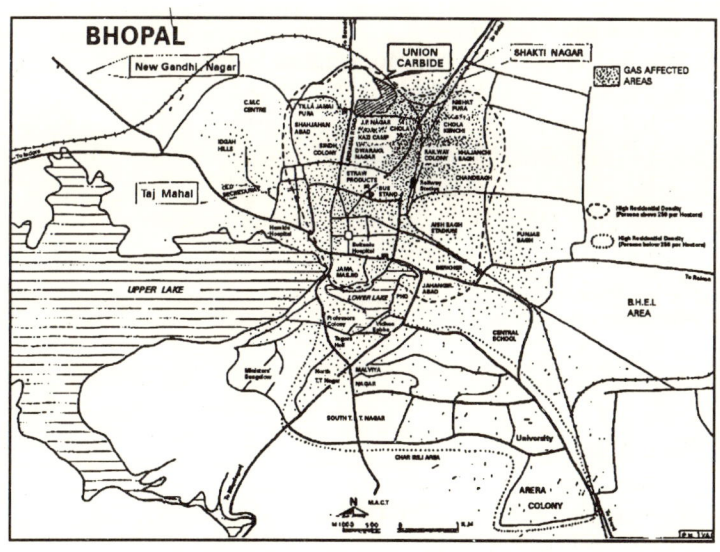

Figure 4   *Map of distribution of the gases*

12.2 Appendix 2: Possible Components of the Gas Cloud

12.2.1 Possible components of the gas cloud

- Alkylamines
- Carbon dioxide CO_2
- Carbon monoxide CO
- Carbon tetrachloride (CCl_4)
- Chloride
- Chloroform
- Di-methylamine hydrochloride
- 1,3-Dimethylisocyanurate (DMI)
- Dimethyl urea (DMU)
- Dione
- Hydrogen H_2
- Hydrogen chloride HCl
- Hydrogen cyanide HCN
- Hydrolyzable chloride
- Hydroxy radical OH-
- Metallic ions (Fe, Cr, Ni, Mo, Na, Ca, Mg)
- Methylcarbamoyl chloride
- Methylchloride
- Methylen dichloride (CH_2Cl_2)
- Methyl isocyaniad (MIC)
- MIC-trimer (MICT)
- Monomethylamine hydrochloride
- Nitric acid
- Nitric oxide NO
- Nitrogen N_2
- Nitrous oxides (NO_x)
- Oxygen O_2
- Phosgene
- Tetramethyl biuret (TRMB)
- Tri-methylamine hydrochloride
- Trimethyl biuret (TMB)
- 1,3,5-trimethyl isocyanurat
- Trimethyl urea (TMU)

- Water H_2O
- Unidentified tarry materials
- Unidentified compounds in the category of molecular weight of 269, 279 etc.

12.2.2 Components found in solid samples from tank 610

- MIC-trimer (MICT) (40-55%),
- 1,3-Dimethylisocyanurate (DMI) (13-20%)
- Dimethyl urea (DMU)
- Trimethyl urea (TMU)
- Trimethyl biuret (TMB)
- Tetramethyl biuret (TRMB)
- Chloride
- Tri-methylamine hydrochloride (3-4%)
- Di-methylamine hydrochloride (2-2,5%)
- Monomethylamine hydrochloride (1-1,5%)
- Hydrolyzable chloride (2-3%)
- Dione
- Metallic ions (Fe, Cr, Ni, Mo, Na, Ca, Mg)
- Unidentified tarry materials (3-6%)
- Several other unidentified compounds in the category of molecular weight of 269, 279 etc.

12.2.3 List of components from the manual of UCC

The composition of combustion gases from a pool of burning MIC is dependent on the assumptions about the air (601). e.g:
- Nitrogen N_2 70%
- Carbon dioxide CO_2 14%
- Water H_2O 14%
- Carbon monoxide CO 1.5%
- Oxygen O_2
- Hydrogen H_2
- Hydroxy radical OH-
- Nitric oxide NO

12.2.4 Possible components according to the CSIR-report

- MIC
- Carbon dioxide
- Methylene dichloride
- Carbon tetrachloride
- Alkylamines
- Phosgene
- Methylcarbamoyl chloride
- Hydrogen chloride

12.3 Appendix 3: UCC Accidents and Pollution

12.3.1 Accidents in the UCIL plant in Bhopal

It said that 19 accidents happened in the plant during 1976–1984. These are documented.

Table 8 *Accidents in the UCIL plant in Bhopal*

Ref	Year	Nature of accident	Injuries
1	1976	Five serious accidents	1 blind 1 chemical burns
1, 2	1978	Fire in naptha stores	Property worth Rs 6.2 millions (US$ 5 million) damaged
1–3	1981	Phosgene spilled on a worker who removed his face mask	1 worker killed 2 workers injuried
1–4	1982 Jan	Phosgene leakage	24 or 28 workers injuried 16 workers admitted to hospital
5	1982 Feb	MIC leak	18 workers
1	1982 Aug	Spill of liquid MIC	1 chemical engineer 30% burn injuries
2	1982	Short circuit in electical panel	3 workers burn injuries 15 workers eye irritation
1, 2, 4	1982 Oct	Leak of MIC, methyl-carbaryl chloride, chloroform and hydrochloric acid from a broken valve	1 supervisor chemical burns 2 or 3 workers severely exposed Residents experienced burning

Continues...

...*Continued*

Ref	Year	Nature of accident	Injuries
			in eyes and breathing trouble and ran away in panic
1	1983	Two MIC leakages	
4	1983–1984	Several leaks of MIC, chlorine, monomethyl-amine, phosgene, and carbon tetrachloride, sometimes in combinations	
2	1986	Leak of chloroform, MIC and hydrochloric acid	
2	1984	Using hands instead of tool	1 worker fracture

12.3.2 Chemical waste dumped by UCIL in and around the factory

Source: Testimonies of ex-UCIL workers, and CSIR and IICT findings submitted in USA and in the Supreme Court of India [6, 7].

Table 9   Chemical waste dumped by UCIL in and around the factory

S. No.	Chemical	Amount	Use in factory	Nature of original pollution
1	Methylene chloride	100 MT	Solvent	Air
2	Methanol	50 MT	Solvent	Air
3	Ortho-dichlorobenzene	500 MT	Solvent	Air, Water, Soil
4	Carbon tetrachloride	500 MT	Solvent	Air
5	Chloroform	300 MT	Solvent	Air

Continues...

...Continued

S. No.	Chemical	Amount	Use in factory	Nature of original pollution
6	Tri methylamine	50 MT	Catalyst	Air
7	Chloro benzyl chloride	10 MT	Ingredient	Air, Water, Soil
8	Mono chloro toluene	10 MT	Ingredient	Air, Water, Soil
9	Toluene	20 MT	Ingredient	Air, Water, Soil
10	Aldicarb	2 MT	Product	Air, Water, Soil
11	Carbaryl	50 MT	Product	Air, Water, Soil
12	Benzene Hexachloride	5 MT	Ingredient	Air, Water, Soil
13	Mercury	1 MT	—	Water, Soil
14	Mono methyl amine	25 MT	Ingredient	Air
15	Chlorine	20 MT	Ingredient	Air
16	Phosgene	5 MT	Ingredient	Air
17	Hydro chloric acid	50 MT	Ingredient	Air, Soil
18	Chloro sulphonic acid	50 MT	Ingredient	Air, Soil
19	Alpha Naphthol*	50 MT	Ingredient	Air, Soil
20	Napthalin	50 MT	Ingredient	Air
21	Chemical waste Tar	50 MT	Waste	Water, Soil
22	Methyl Iso cyanate	5 MT	Ingredient	Air, Water, Soil

* During the unsuccessful operation of the alpha-naphthol plant, several chemical compounds weighing over 100 MT also caused pollution of soil, water and air.

12.3.3 Accidents in UCC plants except Bhopal

From the UCC recearch compendium [8, 9], we get the following figures:

In February, 1985, UC notified the US EPA that since 1980, chemical leaks had occurred at UCC's Institute, West Virginia, including 68 leaks of MIC, 196 of phosgene, and 22 of both.

The compendium includes a list of chemical spills and leaks in UC plants in USA 1980–1994. It includes nearly 600 occasions.

In 1986 US OSHA (Occupational Safety & Health Administration), after a September 1985 inspection of five of 18 plant units at Institute, West Virginia, alleged 221 violations of 55 health and safety laws. Of these, 72 were classified as "serious", that is, risk of substantial probability of death or substantial physical harm.

In 1993, Union Carbide's Responsible Care Progress Report listed the number of fires and explosions per year as 6 in 1988, 3 in 1989, 3 in 1990, 2 in 1991 and 2 in 1992 [10].

Some of the most dangerous accidents are listed in table 10 given on pp. 267–271.

A more realistic account of these accidental releases is provided by the table 11 which lists numbers of spills by Union Carbide in the US from 1985 to August 1994 [10].

Table 11   *Numbers of spills by Union Carbide in the US*

Year	Number of spills
1985	8
1986	3
1987	12
1988	10
1989	14
1990	95
1991	134
1992	120
1993	89
1994	94

12.3.4 Toxic releases and waste generation in USA

Mac Sheoin [10] gives a detailed description of UCC's pollution in USA. In 1987, toxic releases were 69,556,343 lbs (31,579 tonnes), compared to 6,063,839 lbs (2,753 tonnes) in 1999. The amount of waste did not decrease in the same way: 185,494,608 lbs (84,214 tonnes) in 1991 and 164,247,503 lbs (74,568 tonnes) in 1999.

Table 10 Most dangerous accidents in UCC plants except Bhopal

Ref			Disease	Injured	Killed	Society
5, 11	1930–31	Gauley Bridge, West Virginia	Tunnel construction in a rock of almost pure silica, without any safety precautions	Silicosis	5000? (total amount of employed workers)	476 workers immediately, several later
9	1950s–1960s	Oak Ridge, Tennessee	Separation of lithium-6 for mercury production. 2.4 million pounds of mercury are unaccounted for. 475,000 pounds leaked into creek. 30,000 pounds are believed to have leaked into the air.	Mercury levels in urine above those considered safe.		Half of workers
10	1969	Texas City	Explosion and fire				Damaged nearby houses
5	1970	Tonawanda, New York	An 18-worker department making molecular sieves	Emphysema	7 out of 18 workers	
10	1973	Penuelas, Puerto Rico				3 workers	
10	1973	Penuelas, Puerto Rico	Benzene leak			1 worker	
10	1973	Institute, West Virginia	Propane fumes			1 worker	

Continues...

...Continued

Ref			Disease	Injured	Killed	Society
10	1975	Antwerp, Belgium	Explosion at PE-plant		6 workers	
[5]	1978		NIAX catalyst ESN, used in the production of polyurethane foam			
10, 12	1978	Near Jakarta, Indonesia	Cimanggis Battery plant	Bladder paralysis in workers		
5, 12?	1978	Near Jakarta, Indonesia	Cimanggis Battery plant	Electrocuted when depressed an unshielded electrical switch		
10	1979	Beauharnois, Quebec	Strike-bound ferromanganese plant	Cadmium poisoning of workers	402 out of 750 workers had kidney diseases	Contamination of surroundings
13	1980	Texas City	Reported second-highest rate of brain cancer of 7 petro-chemical plants	Brain cancer		5 supervisors
9, 10	1980 or 1982	South Charleston, West Virginia	Hydrogen chloride leaked from a tankcar		18 workers	
				A few		Hundreds of residents were evacuated

Continues...

...*Continued*

Ref		Disease	Injured	Killed	Society
11	1981 and before	Paducah, Kentucky	Uranium enrichment plant	Radiation exposure	?		17,000 people were evacuated. Windows blown out.
10	1982	Taft, Louisiana	Tank containing acrolein exploded				
	1984	**Bhopal, India**	**Pesticide plant. Water entered a tank containing 73 tonnes of MIC. A toxic cloud spread over the town.**	Respiratory organs, eyes, neurotoxic, gastro-intestinal, mental	100,000 – 200,000	14,000? First weeks, 15,000? since then	520,000 exposed.
5, 8	1985	Institute, West Virginia	Pesticide plant. After installing a new safety system, MIC and aldocarboxime, used for Temik, leaked. The safety system was not programmed properly.		135, including 6 workers		Thousands of inhabitants exposed. Food contamination in surroundings
10	1985	Taft, Louisiana	Sharp rise in temperature in a peracetic acid and ethyl acetate storage tank				400 people evacuated

Continues...

...*Continued*

Ref			Disease	Injured	Killed	Society
10	1985	Taft, Louisiana	Explosion and fire in the ethylene unit	Dozens of residents		
10	1985	South Charleston, West Virginia	5,700 pounds of acetone and mesityl oxide leaked from a distillation column	Dozens of residents		
9	1985	South Charleston, West Virginia	Hydrochloric acid leak			
9	1985	South Charleston, West Virginia	Dimethyl disulfide leak at Larvin pesticide Unit			
10	1985	Institute, West Virginia	Methylene chloride and aldicarb oxime leaked	6 workers, over a hundred residents		
10	1986	Institute, West Virginia	Explosion	1 worker		
10	1988	Institute, West Virginia	Fire and explosion of 4,300 pounds of ethylene oxide			

Continues...

...Continued

Ref				Disease	Injured	Killed	Society
8	1990	Institute, West Virginia	Leak of MIC and muriatic acid		7 workers		15,000 residents ordered indoors
8	1990	Institute, West Virginia	Two more leaks				
10	1991	Seadrift, Texas	Explosion		1 contract worker	26 workers	
10	1991	Seadrift, Texas	Ammonia spill			8 contract workers	
10	1991	South Charleston, West Virginia	Ethylene oxide leak			2 workers	
10	1993	South Charleston, West Virginia	Anhydrous ammonia leak			1 person	

12.3.5 References

1. Morehouse, W., *The Ethics of Industrial Disaster in a Transnational World: The elusive quest for justice and accountability in Bhopal.*
2. Chauhan, P., S., *Bhopal Tragedy. Socio-legal Implications.* 1996, Rawat Publications: Jaipur.
3. Cassels, J., *The Uncertain Promise of Law: Lessons from Bhopal.* 1993, Toronto: University of Toronto Press Inc.
4. Chouhan, T.R., *Bhopal: The Inside Story. Carbide workers speak out on the world's worst industrial disaster.* 1994, New York: The Apex Press.
5. *The Trade Union Report on Bhopal.* 1985, ICFTU–ICEF: Geneva, Switzerland.
6. *Foul Debris. The UCIL plant is still a health hazard.* Down To Earth, Dec 15, 2003.
7. *Union Carbide's Factory in Bhopal: Still a Potential Killer.* 2002, Bhopal International Coalition for Justice in Bhopal www.bhopal.net.
8. *Contempt of People. Ramifications of the Bhopal Gas Leak Disaster Case.* 1992, Delhi Science Forum: New Delhi.
9. *Union Carbide Corporation's Toxic Chemical Spills.* 2002, Bhopal International Coalition for Justice in Bhopal www.bhopal.net.
10. Mac Sheoin, T., *Report on Union Carbide Corporation.* 2002, The Other Media: New Delhi.
11. Jones, T., *Corporate Killing. Bhopals Will Happen.* 1988, London: Free Association Books.
12. Weir, D., *The Bhopal Syndrome. Pesticides, Environment and Health.* 1988, London: Earthscan publications Limited.
13. *Chronology of the Union Carbide Corporation.* 2002, Bhopal International Coalition for Justice in Bhopal www.bhopal.net.

12.4 Appendix 4: Non-Governmental Organisations and Networks engaged in Bhopal

12.4.1 Local engagements

12.4.1.1 Immediate engagement

- Mahila Chetna Samaj
- Self-Employed Women's Association
- *Indian Red Cross Society*
- *Institute of Engineering of Ahmedabad*
- The Muslim Welfare Society
- *The Lions Club*
- *Jan Sewashya Kendra*
- *Sultan-ul-Hind Mission*
- *Ramakrishna Misson*
- *Mother Teresa's Missionaries of Charity*
- Eklavya *A local environmental organisation.*
- *Nagrik Rahat aur Punawas Committee* Organised opposition.
- *JSK (People's Health Center)* A clinic set up by combined opposition forces in June 1985. It was a free out-patient clinic, working with education and public health. Because of overloading and troubles with the authorities, it was closed down after a few years.
- *The Royal Commonwealth Society for the Prevention of Blindness* Provided an eye hospital.

12.4.1.2 Long term engagement

- *Bhopal Gas Peedit Mahila-Stationery Karmachari Sangh* A trade union formed by the employed women in the only surviving workshed for gas victims.
- *Bhopal Gas Peedit Mahila Udyog Sangathan* The largest local organisation, with several thousands of female members. The male leader Mr Abdul Jabbar Khan keeps the survivor issues alive through regular meetings and contact with media and authorities.

- *Nirashrit Pension Bhogi Sangharsh Morcha* An organisation for pensioners, now less active.
- *Zahreeli Gas Kand Sangharsh Morcha (Poisonous Gas Incident Struggle Front) ("the Morcha")* The organisation was formed immediately after the disaster. Working for empowerment of women, including education. Probably no activities at the moment.
- *Nagrik Rahat aur Punarwas Samiti (Citizens' Relief and Rehabilitation Committee) (NRPS)*
- *Action for Gas Affected People (AGAPE)* Put up a small dispensary.
- *Jan Swasthya Samiti (JSS)* Put up a health dispensary.
- *Bhopal Relief Trust* Study on the health situation.
- *Bhopal Group for Information and Action* Consisting of a few activists with a large international network. Coordinating activities like the Permanent Peoples' Tribunal and the International Medical Commission on Bhopal.
- *Children against Carbide*
- *Bhopal People's Health and Documentation Clinic (Sambhavna Clinic)* A clinic for the worst affected survivors. Opened in 1996, economic support from the Sambhavna Trust. Here is provided medical care through modern (allopathic) medicine as well as Ayurveda and Yoga. A library on documentation has been built up. Sambhavna Trust is currently the only non-governmental medical initiative in Bhopal for the welfare of the survivors.

12.4.2 National (Indian) engagements

- *Medico Friend Circle* Women with medical/non-medical backgrounds who want to change the health system to favour the poor. Studies on women's reproductive health. Provided medical staff to JSK (the People's Health Center).
- *Drug Action Forum* Provided medical staff to JSK.
- *Saheli*, New Delhi. A women's group.
- *Delhi Science Forum*. Filed a detailed submisison with the Commission of Inquiry of the Government of Madhya Pradesh.
- *Bhopal Gas Peedit Sangharsh Sahayog Samiti (BGPSSS)*. Connections with Delhi Science Forum. Active in the criminal case.

- *The Other Media*, New Delhi. Part of the network working on Bhopal. Administered the Fact Finding Mission on Bhopal, 2001–2002.
- *Rashtriya Abhiyan Samiti (National Campaign Committee) (RAWS)* Organised national protests and supported groups in Bhopal politically and financially.
- *Kerala Sastra Sahitya Parishad (KSSP)*, Thiruvananthapuram
- *Union Research Group (URG)*, Mumbai. Has established a Trade Union Relief Fund.
- *Lawyers Collective*, Mumbai Looking into the legal issues.
- *People's Union for Civil Liberties* Made an investigation on the demolition of houses in Bhopal 1991.
- *Society for Participatory Research in Asia (PRIA)*, New Delhi. Published a booklet on Bhopal in Hindi.
- *Center for Science and the Environment*
- *The Sambhavna Trust* A charitable trust registered in June 1995 for the welfare of the survivors through medical care, research, education and information dissemination. Funds are being raised through individual contributions, mainly through campaigns in UK and USA.
- *Indian Law Institute*, New Delhi
- *Bombay Urban Industrial League for Development (BUILD) Documentation Centre*

12.4.3 National engagements outside India

12.4.3.1 North America

- *Participatory Research Group (PRG)*, Toronto
- *Bhopal Action Resource Center*, New York. Mr Ward Morehouse is a well known and most active person organising the American campaign for collecting funds for Sambhavna (Bhopal Medical Appeal).
- *Highlander Research and Education Center*, Tennessee. Provided PRIA with information directly after the Bhopal disaster.
- *Citizen Action Group*, West Virginia
- *People concerned about MIC, Institute*, West Virginia. Monitors the situation at Virginia.
- *Southside Concerned Citizens*, Chatham, Virginia

- *Workers' Policy Project of New York*
- *National Bhopal Disaster Relief Organization.* Collected money.
- *Appalachian Ohio Public Interest Campaign (AOPIC)*, Athens, Ohio
- *Environmental Policy Institute,* Washington, DC
- *National Campaign Against the Misuse of Pesticides (NCAMP)*, Washington, DC
- *Pesticide Action and Education Project,* San Fransisco, California
- *International Institute of Concern for Public Health,* Canada. Main organiser of IMCB.
- *Council on International and Public Affairs,* New York. Mr Ward Morehouse's organisation.
- *National Toxics Campaign Fund*
- *Voluntary Health Association of India*
- US Citizens' Commission on Bhopal
- *Center for Investigative Reporting,* London

12.4.3.2 South America

- *Consejo de Educacion de Adultos de America Latina (CEEAL)*, Santiago, Chile
- *Mision Industriel de Puerto Rico,* Hato Rey, Puerto Rico

12.4.3.3 Asia

- *Association for the Rights of Industrial Accident Victims (ARIAV)*, Hong Kong
- *Asian Pacific Peoples' Environment Network (APPEN)*
- *Bhopal Disaster Monitoring Group,* Tokyo
- *Bhopal Information Network Japan,* Tokyo
- *Campaign against foreign domination in New Zealand,* Christchurch, NZ
- *Asian Victims for a Hazard Free Environment,* Hong Kong. Consists of two parts, ARENA and DAGA.
- *Asian Regional Exchange for New Alternatives (ARENA),* Hong Kong
- *Arena Press,* Hong Kong
- DAGA, Hong Kong

12.4.3.4 Europe

- *ICAS,* Antwerp

- *Wereld Solidariteit,* Brussels
- *British Group for Justice in Bhopal,* London. Connected with the Pesticide Trust. Organising the English campaign for collecting funds for Sambhavna.
- *The Pesticide Trust,* London
- *Bhopal Action Group/TICL,* London
- *Bhopal support network,* Ireland
- *Bhopal Never Again Action Group,* Utrecht
- *Participatory Research Network,* Carnfield
- *Livingston Action Committee for an Inquiry into the Siting of Union Carbide,* Scotland
- *Transnationals Information Centre,* London

12.4.3.5 *Africa*

Environmental Liaison Centre, Nairobi

12.4.4 *International networks*

- *The "No-More-Bhopals" network*
- *International Participatory Research Network*
- *International Organization of Consumer Unions (IOCU), Penang, Malaysia*
- *Bhopal Medical Appeal* The network that organises the campaigns to collect money to Sambhavna.
- *International Coalition for Justice in Bhopal (ICJB)* This is a coalition of organisations representing survivors groups and their supporters. Sent petition to Permanent Peoples' Tribunal and arranges the international protests towards Union Carbide and Dow.
  - Bhopal Action Resource center, USA
  - Bhopal Information Network, Japan
  - Center for Health & Environment, USA
  - Corpwatch, India
  - Essential Action, USA
  - Ecology Centre of Michigan, USA
  - Environmental Health Fund, USA
  - Environmental Health Watch, USA
  - Greenpeace International

- National Campaign for Justice in Bhopal, India
- Pesticide Action Network (PEN), North America
- The Other Media, India
- UK Campaign for Justice in Bhopal, UK
- *International Network for Victims of Corporate and Government Abuse* Supported Permanent Peoples' Tribunal Asia'92.
- *The Pesticide Action Network* (PAN)
- *Asia '92: Permanent Peoples' Tribunal 1992* The Permanent Peoples' Tribunal (PPT) is the successor to the Bertrand Russell Tribunals on crimes against humanity in Vietnam and Chile. The Tribunal attempts to fill gaps in international law, which it seeks to influence by elaborating on such documents as the Universal Declaration of Rights of Peoples (the Algiers Declaration), the Universal Declaration of Human Rights (United Nations), the Nürnberg principles, United Nations General Assembly resolutions on de-colonizations and the new international economic order. During 1991–1994, the PPT had four hearings in relation to the Bhopal Gas disaster. "The Tribunal regards the anti-humanitarian effects of industrial and environmental hazards not as an unavoidable part of the existing industrial system, but rather as a pervasive and organised violation of the most fundamental rights of humanity." Some of the conclusions are listed below. Creation of an independent international Medical Commission on Bhopal was recommended. In 1996, the Charter on Industrial Hazards and Human Rights was drawn up.
  - The fundamental human rights of the victims have been grossly violated, by Union Carbide and by the Government of India.
  - Existing mechanisms to deal with the consequences of such catastrophes, and to prevent them happening in the first place, have failed miserably thus far in Bhopal. Equally conspicuous for their failure are the two prevailing modes for dealing with threats to the environment and human safety, namely, industry self-regulation and government policing.
  - Also conspicuous in their failure to help ameliorate the distress of the victims have been, with some notable exceptions, the legal and medical professions.

- Under no circumstances can the killing and injury of so many innocent people be considered an acceptable cost of developoment.
- *The International Medical Commission on Bhopal 1994 (IMCB)* As a follow-up of the PPT, the IMCB, consisting of 13 medical doctors from 12 countries and one engineer, met in January 1994. Hearings and press conferences were held. Seven articles were published and presented in Bhopal 1996. The recommendations were on social, occupational, economic and environmental rehabilitation, and on developing primary healthcare. The commission offered the government their expertise in planning primary healthcare, without any result so far. IMCB is now dissolved, but individual members still have contact with the government, activists and survivors.

12.5 Appendix 5: Demands from Non Governmental Organisations

12.5.1 Unresolved medical issues of the Union Carbide disaster in Bhopal [1]

1. Research and monitoring
    i. Absence of research on gynaecological, musculosceletal, cardiovascular, renal, gastrointestinal and endocrine system injuries attributable to exposure.
    ii. Absence of administrative set-up for carrying out long term, and possibly trans-generational research activities in · Bhopal.
    iii. Absence of epidemiological and clinical research on continuing exposure related mortality and morbidity.

2. Information
    i. Information on health impacts not available with government and private doctors involved with the medical care of the survivors.
    ii. No official initiative towards disseminating information on health impacts and preventive and ameliorative measures to the survivors.
    iii. Reports on studies carried out by ICMR remain to be published for years after termination of research.

3. Health care
    i. Absence of perspective and administrative set-up to respond to the chronic nature of exposure related diseases.
    ii. Absence of protocols for treatment of exposure induced illnesses.
    iii. Absence of a system of recording health status and efficacy of medical interventions.
    iv. Absence of a community based health care approach and overwhelming emphasis on hospital based treatment.
    v. Negligence of indigenous systems of medicine.
    vi. Unavailability of medicines and facilities for investigations at government hospitals and clinics.

vii. Indiscriminate use of steroids, antibiotics, psychotropic and symptomatic drugs.

viii. Absence of drug-free therapies, such as Yoga.

4. Health education and public health improvement

    i.   Absence of official initiatives towards health education among survivors.

    ii. No official initiatives towards provision of clean air, water and sanitation facilities to the survivors.

    iii. No official action on toxic contamination of groundwater and soil in the communities adjacent to the Union Carbide factory.

12.5.2 Survivors' demands [2]

From Dow Chemical

1. That Dow Chemical releases vital medical information on composition of gases and its effect on humans.
2. Provide complete medical, social and economic rehabilitation.
3. Clean up the contaminated factory site including groundwater supplies.
4. Dow officials, including ex-Carbide CEO Anderson, appear in Indian courts to face criminal charges of *'culpable homicide'*.

From the Indian government

5. Distribute the remaining compensation funds to the affected population including orphaned and unregistered children.
6. That the Indian Council for Medical Research (ICMR) releases findings from all the 24 studies that were conducted on gas affected persons and were mysteriously shut down by 1992.
7. Scrap the illegal and arbitrary announcement by Welfare Commissioner, Bhopal Gas Victims, that wrongly denies compensation to over 10,000 victims.
8. Support the Class Action suit filed by the survivors against Carbide and Warren Anderson in the US.

9. Take action against Union Carbide and Bhopal's Municipal Corporation for unauthorised withdrawal and expenditure of funds meant for Bhopal victims.
10. Set up regulatory mechanisms to ensure that reckless medication at the Bhopal Memorial Trust is stopped.

12.5.3 The International Campaign for Justice in Bhopal [3]

1. Extradite Warren Anderson, ex-Chairman of Union Carbide to face criminal charges in the Indian court at which he has been thumbing his nose for the last 11 years.
2. Secure fair and just compensation for the victims of the 1984 disaster who have had to struggle against illness and destitution for more than 18 years on "compensation" that works out at the price of one cup of tea per day.
3. Ensure that the balance of the fund meant for the welfare of the victims and currently being withheld by the Government of India is distributed to the victims, and is not diverted to any other purpose.
4. Compel Dow Chemical, the 100% owner of Union Carbide, to clean up its subsidiary's abandoned factory in Bhopal in accordance with the universally-acknowledged "polluter pays" principle.
5. Secure just compensation for those who have been injured or made ill by the ongoing contamination at Union Carbide's derelict factory, which continues to poison soil and drinking water supplies.
6. Compel Dow Chemical, on behalf of its subsidiary Union Carbide, to provide for the long-term health care of the survivors of the 1984 disaster and of the ongoing contamination, including care for future generations who face risks that only now are becoming apparent.
7. Compel Dow Chemical to make restitution for loss of earnings by victims of the 1984 gas leak and the ongoing contamination.
8. To give support and solidarity to other organisations and campaigns fighting against toxic terror and corporate crime.

9. To educate people about the devastating long term effects of toxic chemicals on health and to expose the risks that communities all over the world, including the USA, face daily.

12.5.4 References

1. *The Bhopal Gas Tragedy 1984– ? A report from the Sambhavna Trust, Bhopal, India.* 1998, Bhopal People's Health and Documentation Clinic: Bhopal.
2. *Hinterland. A Special Issue on the Bhopal Gas Tragedy.* 2003, Department of English, Hindu College: New Delhi. p. 34.
3. *International Campaign for Justice in Bhopal.* www.bhopal.net. 23 November 2002.

Related titles

Agriculture and Intellectual Property Rights: Economic, Institutional and Implementation Issues in Biotechnology	*Santaniello, V. et. al.*
Battles over Nature: Science and the Politics of Conservation	*Saberwal, Vasant and Rangarajan, Mahesh*
Biotechnological Methods of Pollution Control	*Abbasi, S. A. & Ramasami, E. V.*
Coping with Natural Hazards: Indian Context	*Valdiya, K. S.*
Corporate Environmental Management	*Welford, R. et. al.*
Disaster Management	*Harsh K. Gupta (ed.)*
Earth Policy Reader, The	*Brown, Lester, Larsen, Janet & Fischlowitz-Roberts Bernie*
Eco-Economy: Building an Economy for the Earth	*Brown, Lester, R.*
Geology, Environment and Society	*Valdiya, K. S.*
Global Warming: Can Civilization Survive?	*Brown, P.*
Green, Inc.: A Guide to Business and the Environment	*Cairncross, F.*
Greenhouse: The 200-year Story of Global Warming	*Christianson, G. E.*
Intellectual Property Rights in Agricultural Biotechnology	*Erbisch, F. H. & Maredis, K. M. (Eds.)*
Power Play: A study of the Enron Project	*Mehta, Abhay*
Reader in Business and the Environment	*Welford, R. & Starkey, R.*